Who's #1?
The Science of Rating and Ranking

Who's #1?
The Science of Rating and Ranking

Amy N. Langville and Carl D. Meyer

PRINCETON UNIVERSITY PRESS

PRINCETON AND OXFORD

Published by Princeton University Press
41 William Street, Princeton, New Jersey 08540

In the United Kingdom: Princeton University Press
6 Oxford Street, Woodstock, Oxfordshire, OX20 1TW

Library of Congress Cataloging-in-Publication Data
Langville, Amy N.
Who's #1?: The science of rating and ranking / Amy N. Langville and Carl D. Meyer.
p. cm.
Includes bibliographical references and index.
ISBN 978-0-691-15422-0 (alk. paper)
1. Ranking and selection (Statistics) I. Meyer, C. D. (Carl Dean) II. Title.
QA278.75.L36 2012
519.5–dc23   2011036016

British Library Cataloging-in-Publication Data is available

This book has been composed in LATEX

The publisher would like to acknowledge the authors of this volume for providing the camera-ready
copy from which this book was printed.

Printed on acid-free paper. ∞

press.princeton.edu

Printed in the United States of America

10 9 8 7 6 5 4 3 2 1

To John,

for whom I need no algorithm
to tell me who is, was,
and ever shall be
my #1.

AND

To Bethany,

Rating = 10,   Rank = 1.

Contents

Preface

"Always the beautiful answer who asks a more beautiful question." —e.e. cummings

Purpose

We became intrigued by the power of rating and ranking methods while writing our earlier book *Google's PageRank and Beyond: The Science of Search Engine Rankings* [49]. While developing this "Google book," we came to appreciate how rich (and old) the area of rating and ranking is and how far beyond Web search it extends. But learning about the various facets of the subject was tedious because we could not find a convenient source that neatly tied the multitude of methods and applications together into one nice package. Thus this book was born. In addition to collecting in one place relevant information that is spread throughout many individual articles, websites, and other sources, we also present several new ideas of our own concerning rating and ranking that have heretofore not been published. Our goal is to arm readers with enough background and diversity to generate an appreciation for the general art of rating and ranking as well as to prepare them to explore techniques and applications beyond those that appear in this book.

Audience

As the list on page 6 at the end of Chapter 1 points out, the applications of ranking are extremely numerous and varied. Consequently, there are many types of readers who might be interested in the topic, and likewise, this book. For example, readers might be sports enthusiasts, social choice theorists, mathematicians, computer scientists, engineers, and college and high school teachers of classes such as linear algebra, optimization, mathematical modeling, graph theory, or special topics classes, not to mention the people interested in wagering on just about anything that you can think of.

Prerequisites

Most of the book assumes that the reader has had a course in elementary linear algebra and a few chapters assume some knowledge of optimization. If you haven't had these classes, we have a few suggestions.

1. You should read through the book anyway. Every linear algebra-based method in the book boils down to one of two fundamental elements of the subject: solving a system

of linear equations or computing an eigenvector. Even if you don't understand these two elements, you can still implement the methods. For example, there is software, both free and for a fee, that can solve a system of linear equations or compute an eigenvector for you. You just need to understand what data is required as input and how to interpret the output—an understanding of what your software is doing internally is not required.

2. Of course, a true understanding of these subjects will improve your ability to implement, modify, and adapt the methods for your specific goals, so you might consider self-study with an online tutorial or a classic book. Our favorite linear algebra books are [54, 76, 40], but if you are a complete novice, then seek out an online elementary tutorial (many exist) concerning the topic(s) in question. A few of our favorite optimization books, in increasing level of depth, are: [82, 60, 10, 84, 85].

3. Finally, and ideally, we recommend taking an elementary linear algebra class, and later, if time and money permit, an optimization class.

Teaching from This Book

We anticipate that college or high school teachers will use this book for either a short 1–2 lecture module or an entire class such as a special topics or mathematical modeling class. In the former case, teachers will find it quite easy to extract material for a short module. In fact, we have "modularized" the material with this aspect in mind—i.e., very little in the book builds on prior material. The various chapters can be presented in nearly any order. For example, a linear algebra teacher might want to add a lecture on an application of linear systems. In this case, a good choice is Chapter 3 on the Colley method, which is the simplest, most straightforward linear system method in the book. Similarly, after a lecture on the theory of eigenvalues and eigenvectors, a teacher may choose to lecture on an application of eigensystems using the material in Chapter 4. Chapter 6 contains material on Markov chains, a topic that typically appears toward the end of a linear algebra class.

Acknowledgments

- **The College of Charleston.** Luke Ingram and John McConnell, two former M.S. students, completed a class project on ranking teams in the March Madness basketball tournament in the spring of 2006. Luke Ingram extended this class project and created an outstanding thesis [41] with several interesting ranking ideas. Then in the spring of 2008, two undergraduate mathematics majors, Neil Goodson and Colin Stephenson, tackled the same class project, predicting games in that year's March Madness tournament. Neil and Colin used a very preliminary draft of this book and tested several of the ranking models presented herein. They did such a good job of predicting games in the 2008 tournament that they earned a shocking amount of national press. See the aside and notes on pages 151 and 212. Kathryn Pedings, a B.S. Mathematics graduate, continued on as a M.S student working on her thesis topic of linear ordering. She served as assistant extraordinaire by collecting data used in several asides and participating in several brainstorming and discussion sessions. Much of Kathryn's work appears in Chapters 8 and 15. Finally, all students of the Ranking and Clustering research group at the College of Charleston participated in weekly research discussions that helped to form this book. Thus, the first author makes additional shout-outs to Emmie Douglas, Ibai Basabe, Barbara Ball, Clare Rodgers, Ryan Parker, and Patrick Moran.

- **North Carolina State University.** As part of her thesis *Ranking Theory with Applications to Popular Sports* [34], Anjela Govan contributed to the development of the Offense-Defense rating theory [35] by establishing and formalizing the connections between the OD method and the Sinkhorn–Knopp theory of matrix balancing. In addition, her work on Web scraping and data collection was the basis for a significant number of experiments on a large variety of rating and ranking methods, some of which are included in this book. Evaluating techniques that do not work well is as important as revealing those that do. We limited our book to methods that have significant merit, so many of Anjela's significant contributions are not transparent to the reader, but the book is much better because of her work. Charles D. (Chuck) Wessell, was a graduate student at NC State at the time this book was being written, and is now part of the mathematics faculty at Gettysburg College in Gettysburg, PA. Chuck provided many helpful suggestions, and we are indebted to him for his eagle sharp eyes. His careful reading of the manuscript along with his careful scrutiny of the data (the NFL data in particular) prevented several errors from being printed. In addition, Chuck taught an undergraduate class from the material in our book, and his class-room experiences helped to hone the exposition.

- **Colleagues.** Discussions with David Gleich during his visits to the College of Charleston influenced aspects in Chapters 14 and 16. Timothy Chartier of Davidson College corresponded regularly with the first author, and Tim's projects with his student Erich Kreutzer along with their feedback on early drafts of this book were insightful. In particular, the discussion on ties in Chapter 11 resulted from this collaboration. Kenneth Massey of Carson-Newman College hosts a huge data warehouse and website dedicated to sports ranking. Almost all of the sports examples in this book owe their existence to Dr. Massey and his data. We are grateful for his generosity, attitude, knowledge, and computer expertise.

- **Support.** The work of the first author was supported in part by the National Science Foundation CAREER award CCF-0546622, which helped fund collaborative travel and a mini-sabbatical to complete and revise the manuscript. Furthermore, the first author is grateful for the support and nurturing environment provided by the College of Charleston. Since Day 1, the university, college, and department have been extremely welcoming and supportive of her career and research. In particular, she thanks the departmental chairs, former and current, Deanna Caveny and Bob Mignone, respectively, as well as former Dean Noonan for their creative leadership and warm support.

- **Photographs.** The National Science and Technology Medals Foundation and Ryan K. Morris are acknowledged for their permission to reprint the photograph of K. Arrow on page 4. The photograph of K. Massey on page 9 is reprinted courtesy of Kenneth Massey, and the photograph of J. Keener on page 51 is reprinted courtesy of James Keener. The photograph of J. Kleinberg on page 92 is by Michael Okoniewski. The photograph of A. Govan on page 94 is reprinted courtesy of Anjela Govan. The photographs of Neil Goodson and Colin Stephenson on page 152 are courtesy of the College of Charleston. The photographs of Kathryn Pedings and Yoshitsugu Yamamoto on page 153 are courtesy of Kathryn Pedings and Yoshitsugu Yamamoto.

Who's #1?
The Science of Rating and Ranking

Chapter One
Introduction to Ranking

It was right around the turn of the millennium when we first became involved in the field of ranking items. The now seminal 1998 paper *Anatomy of a Search Engine* used Markov chains, our favorite mathematical tool, to rank webpages. Two little known graduate students had used Markov chains to improve search engine rankings and their method was so successful that it became the foundation for their fledgling company, which was soon to become the extremely well-known search engine Google. The more we read about ranking, the more involved in the field we became. Since those early days, we've written one book *Google's PageRank and Beyond: The Science of Search Engine Rankings* and a few dozen papers on the topic of ranking. In 2009, we even hosted the first annual Ranking and Clustering Conference for the southeastern region of the U.S.

As applied mathematicians, our attraction to the science of ranking was natural, but we also believe ranking has a universal draw for the following reasons.

- The problem of ranking is elegantly simple—arrange a group of items in order of importance—yet some of its solutions can be complicated and full of paradoxes and conundrums. We introduce a few of these intriguing complications later in this chapter as well as throughout the book.

- There is a long and glorious tradition associated with the problem of ranking that dates back at least to the 13th century. The asides sprinkled throughout this book introduce some of the major and more colorful players in the field of ranking.

- It appears that today the tradition of progress on the ranking problem is reaching a peak in terms of activity and interest. One reason for the increased interest comes from today's data collection capabilities. There is no shortage of interesting and real datasets in nearly every field imaginable from the real estate and stock markets to sports and politics to education and psychology. Another reason for the increased interest in ranking relates to a growing cultural trend fostered in many industrialized nations. In America, especially, we are evaluation-obsessed, which thereby makes us ranking-obsessed given the close relationship between ranking and evaluation. On a daily basis you probably have firsthand contact with dozens of rankings. For example, on the morning drive, there is the top-10 list of songs you hear on the radio. Later you receive your quarterly numerical evaluations from your boss or teacher (and leave happy because you scored in the 95th percentile.) During your lunch break, you peek at the sports section of the newspaper and find the rankings of your favorite team. A bit later during your afternoon break, you hop online to check the ranking of your fantasy league team. Meanwhile at home in your mailbox sits the next Netflix movie from your queue, which arrived courtesy of the postal system and the company's sophisticated ranking algorithm. Last and probably most frequent are the many times a day that you let a search engine help you find information, buy products, or download files. Any time you use one of these virtual machines, you are relying on advanced ranking techniques. Ranking is so pervasive that the topic even appeared in a recent xkcd comic.

Figure 1.1 xkcd comic pertaining to ranking

Several chapters of this book are devoted to explaining about a dozen of the most popular ranking techniques underlying some common technologies of our day. However, we note that the ranking methods we describe here are largely based in matrix analysis or optimization, our specialties. Of course, there are plenty of ranking methods from other specialties such as statistics, game theory, economics, etc.

- Netflix, the online movie rental company that we referred to above, places such importance on their ability to accurately rank movies for their users that in 2007 they hosted a competition with one of the largest purses ever offered in our capitalistic age. Their Netflix Prize awarded $1 million to the individual or group that improved their recommendation by 10%. The contest captivated us as it did many others. In fact, we applauded the spirit of their competition for it haled of the imperial and papal commissions from a much earlier time. Some of the greatest art of Italy was commissioned after an artist outdid his competition with a winning proposal to a sponsored contest. For example, the famous baptistery doors on the Cathedral in Florence were done by Lorenzo Ghiberti after he won a contest in 1401 that was sponsored by the Wool Merchants' Guild. It is nice to see that the mathematical problem of ranking has inspired a return to the contest challenge, wherein any humble aspiring scientist has a chance to compete and possibly win. It turns out that the Netflix Prize was awarded to a team of top-tier scientists from BelKor (page 134). Much more of the Netflix story is told in the asides in this book.

- Even non-scientists have an innate connection to ranking for humans seem to be psychologically wired for the comparisons from which rankings are built. It is well-known that humans have a hard time ranking any set of items greater than size 5. Yet, on the other hand, we are particularly adept at pair-wise comparisons. In his bestseller *Blink*, Malcolm Gladwell makes the argument that snap judgments made in the blink of an eye often make the difference between life and death. Evolution has recognized this pattern and rewarded those who make quick comparisons. In fact, such comparisons occur dozens of times a day as we compare ourselves to others (or former versions of ourselves). I look younger than her. She's taller than I am. Jim's faster than Bill. I feel thinner today. Such pair-wise comparisons are at the heart of this book because every ranking method begins with pair-wise comparison data.

- Lastly, while this final fact may be of interest to just a few of you, it is possible now to get a Ph.D. in ranking sports. In fact, one of our students, Anjela Govan, recently did just that, graduating from N. C. State University with her Ph.D. dissertation on *Ranking Theory with Application to Popular Sports*. Well, she actually passed the final oral exam on the merit of the mathematical analysis she conducted on several algorithms for ranking football teams, but it still opens the door for those extreme sports fans who love their sport as much as they love data and science.

Social Choice and Arrow's Impossibility Theorem

This book deals with ranking items and the classes of items can be as diverse as sports teams, webpages, and political candidates. This last class of items invites us to explore the field of social choice, which studies political voting systems. Throughout history, humans have been captivated by voting systems. Early democracies, such as Greece, used the standard plurality voting system, in which each voter submits just one lone vote for their top choice. The advantages and disadvantages of this simple standard have since been debated. In fact, French mathematicians Jean-Charles de Borda and Marie Jean Antoine Nicolas Caritat, the marquis de Condorcet, were both rather vocal about their objections to plurality voting, each proposing their own new methods, both of which are discussed later in this book—in particular see the aside concerning BCS rankings on page 17 and the discussion on page 165.

Plurality voting is contrasted with preference list voting in which each voter submits a *preference list* that places candidates in a ranked order. In this case, each voter creates a rank-ordered list of the candidates. These ranked lists are somehow aggregated (see Chapters 14 and 15 for more on rank aggregation) to determine the overall winner. Preferential voting is practiced in some countries, for instance, Australia.

Democratic societies have long searched for a perfect voting system. However, it wasn't until the last century that the mathematician and economist Kenneth Arrow thought to change the focus of the search. Rather than asking, "What is the perfect voting system?", Arrow instead asked, "Does a perfect voting system exist?"

Arrow's Impossibility Theorem

The title of this section foreshadows the answer. Arrow found that his question about the existence of a perfect voting system has a negative answer. In 1951, as part of his doctoral dissertation, Kenneth Arrow proved his so-called Impossibility Theorem, which describes the limitations inherent in any voting system. This fascinating theorem states that no voting system with three or more candidates can simultaneously satisfy the following four very common sense criteria [5].

K. Arrow

1. Arrow's first requirement for a voting system demands that every voter be able to rank the candidates in any arrangement he or she desires. For example, it would be unfair that an incumbent should automatically be ranked in the top five. This requirement is often referred to as the *unrestricted domain* criterion.

2. Arrow's second requirement concerns a subset of the candidates. Suppose that within this subset voters always rank candidate A ahead of candidate B. Then this rank order should be maintained when expanding back to the set of all candidates. That is, changes in the order of candidates outside the subset should not affect the ranking of A and B relative to each other. This property is called the *independence of irrelevant alternatives*.

3. Arrow's third requirement, called the *Pareto principle*, states that if all voters choose A over B, then a proper voting system should always rank A ahead of B.

4. The fourth and final requirement, which is called *non-dictatorship*, states that no single voter should have disproportionate control over an election. More precisely, no voter should have the power to dictate the rankings.

This theorem of Arrow's and the accompanying dissertation from 1951 were judged so valuable that in 1972 Ken Arrow was awarded the Nobel Prize in Economics. While Arrow's four criteria seem obvious or self-evident, his result certainly is not. He proves that it is impossible for any voting system to satisfy all four common sense criteria simultaneously. Of course, this includes all existing voting systems as well as any clever new systems that have yet to be proposed. As a result, the Impossibility Theorem forces us to have realistic expectations about our voting systems, and this includes the ranking systems presented in this book. Later we also argue that some of Arrow's requirements are less pertinent in certain ranking settings and thus, violating an Arrow criterion carries little or no implications in such settings.

Small Running Example

This book presents a handful of methods for ranking items in a collection. Typically, throughout this book, the items are sports teams. Yet every idea in the book extends to any set of items that needs to be ranked. This will become clear from the many interesting non-sports applications that appear in the asides and example sections of every chapter. In order to illustrate the ranking methods, we will use one small example repeatedly. Because sports data is so plentiful and readily available our running example uses data from the 2005 NCAA football season. In particular, it contains an isolated group of Atlantic Coast Conference teams, all of whom played each other, which allows us to create simple traditional rankings based on win-loss records and point differentials. Table 1.1 shows this information for these five teams. Matchups with teams not in Table 1.1 are disregarded.

Table 1.1 Game score data for a small 5-team example

	Duke	Miami	UNC	UVA	VT	Record	Point Differential
Duke		7-52	21-24	7-38	0-45	0-4	-124
Miami	52-7		34-16	25-17	27-7	4-0	91
UNC	24-21	16-34		7-5	3-30	2-2	-40
UVA	38-7	17-25	5-7		14-52	1-3	-17
VT	45-0	7-27	30-3	52-14		3-1	90

This small example also allows us to emphasize, right from the start, a linguistic pet peeve of ours. We distinguish between the words *ranking* and *rating*. Though often used interchangeably, we use these words carefully and distinctly in this book. A *ranking* refers to a rank-ordered list of teams, whereas a *rating* refers to a list of numerical scores, one for each team. For example, from Table 1.1, using the elementary ranking method of win-loss record, we achieve the following ranked list.

$$
\begin{matrix}
\text{Duke} \\
\text{Miami} \\
\text{UNC} \\
\text{UVA} \\
\text{VT}
\end{matrix}
\begin{pmatrix}
5 \\ 1 \\ 3 \\ 4 \\ 2
\end{pmatrix},
$$

meaning that Duke is ranked 5th, Miami, 1st, etc. On the other hand, we could use the point differentials from Table 1.1 to generate the following rating list for these five teams.

$$
\begin{matrix}
\text{Duke} \\
\text{Miami} \\
\text{UNC} \\
\text{UVA} \\
\text{VT}
\end{matrix}
\begin{pmatrix}
-124 \\ 91 \\ -40 \\ -17 \\ 90
\end{pmatrix}.
$$

Sorting a rating list produces a ranking list. Thus, the ranking list, based on point differentials, is

$$
\begin{matrix}
\text{Duke} \\
\text{Miami} \\
\text{UNC} \\
\text{UVA} \\
\text{VT}
\end{matrix}
\begin{pmatrix}
5 \\ 1 \\ 4 \\ 3 \\ 2
\end{pmatrix},
$$

which differs only slightly from the ranking list produced by win-loss records. Just how much these two differ is a tricky question and one we postpone until much later in the book. In fact, the difference between two ranked lists is actually a topic to which we devote an entire chapter, Chapter 16.

In closing, we note that every rating list creates a ranking list, but not vice versa. Further, a ranking list of length n is a permutation of the integers 1 to n, whereas a rating list contains real (or possibly complex) numbers.

Ranking vs. Rating

A *ranking* of items is a rank-ordered list of the items. Thus, a ranking vector is a permutation of the integers 1 through n.

A *rating* of items assigns a numerical score to each item. A rating list, when sorted, creates a ranking list.

ASIDE: An Unranked List of Famous Rankings

This book is sprinkled with examples taken from a variety of contexts and applications. In these examples, we assemble ratings, and hence rankings, and compare them to some famous standard lists. It is perhaps ironic that we begin this book on ranking with an unranked list. Nevertheless, the following list, which is in no particular order, provides a brief introduction to several standard lists that you will encounter often in this book. This unranked list also demonstrates the range of settings for which rankings are useful.

- PageRank: Google made this ranking famous with the success they have had using their PageRank algorithm to rank webpages [15, 49]. There is much more on this famous ranking in Chapter 6.

- HITS: This algorithm for ranking webpages [45, 49] was invented around the same time as the PageRank algorithm. It is currently part of the search engine Ask.com. Our Offense–Defense (or OD) method described in Chapter 7 was motivated by concepts central to HITS.

- BCS: College football teams are ranked by their BCS rating. BCS (Bowl Championship Series) ratings help determine which teams are invited to which bowl games. Chapters 2 and 3 cover two rating methods used by the BCS.

- RPI: College basketball teams are ranked by their RPI (Rating Percentage Index), which is used to determine which teams are invited to the March Madness tournament.

- Netflix and IMDb: Movie fans can consult the ranked lists of the best movies of all-time to determine movie rental choices. Both Netflix, the novel postal system for renting movies online, and Internet Movie Database (IMDb) provide ranked lists of the highest rated movies. The asides on pages 25 and 95 use the Netflix Prize dataset to rank movies.

- Human Development Index: The United Nations ranks countries by their Human Development Index (HDI), which is a comparative measure of life expectancy, literacy, education, and standard of living. The HDI is used to make decisions regarding aid to under-developed countries.

- College Rankings: *US News* publishes an annual ranking of colleges and universities in the United States. The aside on page 77 contains more on these rankings and their calculation.

- Chess players: FIDE, the international federation for chess, uses the Elo method of Chapter 5 to rank hundreds of thousands of chess players worldwide.

- Food chain: Species of animals feed on and are food for other species. Thus, a directed graph showing the dominance relationships between different species can be built. This graph can then be analyzed to answer questions such as who is top of the food chain or which species is the linchpin of the entire chain.

- Archaeology: The aside on page 192 shows how ranking methods are used to put artifacts in relative order based on their dig depths at multiple locations.

- Legal cases: The trail of citations from case to case to case leaves valuable information that enables a ranking of legal cases. In fact, current legal search engines use this technology to return a ranked list of cases in response to a paralegal's query on a particular topic.

- Social Networks: The link structure in a social network like Facebook can be used to rank members according to their popularity. In this case, the Markov or OD ranking methods of Chapters 6 and 7 are particularly good choices. However, Facebook may have had its origins in the Elo rating system described in Chapter 5—see the aside on page 64. Of course, some networks, such as the links between terrorists coordinating the 9/11 attacks, carry weightier communications. The same techniques that find the most popular Facebook members can be used to identify the key figures, or central hubs, of these networks.

By The Numbers —

$50,000,000 = $ attendance at American college football games in 2010.

— The Line Makers

Chapter Two

Massey's Method

The Bowl Championship Series (BCS) is a rating system for NCAA college football that was designed to determine which teams are invited to play in which bowl games. The BCS has become famous, and perhaps notorious, for the ratings it generates for each team in the NCAA. These ratings are assembled from two sources, humans and computers. Human input comes from the opinions of coaches and media. Computer input comes from six computer and mathematical models—details are given in the aside on page 17. The BCS ratings for the 2001 and 2003 seasons are known as particularly controversial among sports fans and analysts. The flaws in the BCS selection system as opposed to a tournament playoff are familiar to most, including the President of the United States—read the aside on page 19.

Initial Massey Rating Method

K. Massey

In 1997, Kenneth Massey, then an undergraduate at Bluefield College, created a method for ranking college football teams. He wrote about this method, which uses the mathematical theory of least squares, as his honors thesis [52]. In this book we refer to this as the Massey method, despite the fact that there are actually several other methods attributed to Ken Massey. Massey has since become a mathematics professor at Carson-Newman College and today continues to refine his sports ranking models. Professor Massey has created various rating methods, one of which is used by the Bowl Championship Series (or BCS) system to select NCAA football bowl matchups—see the aside on page 17. The Colley method, described in the next chapter, is also one of the six computer rating systems used by the BCS. Because the details of the Massey method that is used by the BCS are vague, we describe the Massey method for which we have complete details, the one from his thesis [52]. Readers curious about Massey's newer, BCS-implemented model can peruse his sports ranking Web site (masseyratings.com) for more information.

Massey's Main Idea

The fundamental philosophy of the Massey least squares method can be summarized with one idealized equation,

$$r_i - r_j = y_k,$$

where y_k is the margin of victory for game k, and r_i and r_j are the ratings of teams i and j, respectively. In other words, the difference in the ratings r_i and r_j of the two teams ideally predicts the margin of victory in a contest between these two teams.

The goal of any rating system is to associate a rating for each team in a league of n teams, where m total league games have been played thus far. Of course, we do not know the ratings r_i for the teams, but we do know who played who and what the margin of victory was. Thus, there is an equation of this form for every game k, which creates a system of m linear equations and n unknowns that can be written as

$$\mathbf{Xr} = \mathbf{y}.$$

Each row of the coefficient matrix $\mathbf{X}$ is nearly all zeros with the exception of a 1 in location i and a -1 in location j, meaning that team i beat team j in that matchup. Thus, $\mathbf{X}_{m \times n}$ is a very sparse matrix. The vector $\mathbf{y}_{m \times 1}$ is the right-hand side vector of margins of victory and $\mathbf{r}_{n \times 1}$ is the vector of unknown ratings. Typically, $m \gg n$, which means the linear system is highly overdetermined and is inconsistent. All hope is not lost as a least squares solution can be obtained from the *normal equations* $\mathbf{X}^\mathbf{T}\mathbf{Xr} = \mathbf{X}^\mathbf{T}\mathbf{y}$. The least squares vector, i.e., the solution to the normal equations, is the best (in terms of minimizing the variance) linear unbiased estimate for the rating vector $\mathbf{r}$ in the original equation $\mathbf{Xr} = \mathbf{y}$ [54].

Massey discovered that due to the structure of $\mathbf{X}$ it is advantageous to use the coefficient matrix $\mathbf{M} = \mathbf{X}^\mathbf{T}\mathbf{X}$ of the normal equations. In fact, $\mathbf{M}$ need not be computed. It can simply be formed using the fact that diagonal element $\mathbf{M}_{ii}$ is the total number of games played by team i and the off-diagonal element $\mathbf{M}_{ij}$, for $i \neq j$, is the negation of the number of games played by team i against team j. In a similarly convenient manner, the right-hand side of the normal equations $\mathbf{X}^\mathbf{T}\mathbf{y}$ can be formed by accumulating point differentials. The ith element of the right-hand side vector $\mathbf{X}^\mathbf{T}\mathbf{y}$ is the sum of the point differentials from every game played by team i that season, and so we define $\mathbf{p} = \mathbf{X}^\mathbf{T}\mathbf{y}$. Thus, the Massey least squares system becomes

$$\mathbf{Mr} = \mathbf{p},$$

where $\mathbf{M}_{n \times n}$ is the Massey matrix described above, $\mathbf{r}_{n \times 1}$ is the vector of unknown ratings, and $\mathbf{p}_{n \times 1}$ is the right-hand side vector of cumulative point differentials.

The Massey matrix $\mathbf{M}$ has a few noteworthy properties. First, $\mathbf{M}$ is much smaller in size than $\mathbf{X}$. In fact, it is a square symmetric matrix of order n. Second, it is a diagonally dominant M-matrix. Third, its rows sum to 0. As a result, the columns of $\mathbf{M}$ are linearly dependent. This causes a small problem as $rank(\mathbf{M}) < n$ and the linear system $\mathbf{Mr} = \mathbf{p}$ does not have a unique solution. Massey's workaround solution to this problem is to replace any row in $\mathbf{M}$ (Massey chooses the last) with a row of all ones and the corresponding entry of $\mathbf{p}$ with a zero. This adds a constraint to the linear system that states that the ratings must sum to 0 and results in a coefficient matrix that is full rank. A similar trick is employed in the direct solution technique for computing the stationary vector of a Markov chain. See Chapter 3 of [75]. The new system with the row adjustment is denoted $\bar{\mathbf{M}}\mathbf{r} = \bar{\mathbf{p}}$.

The Running Example Using the Massey Rating Method

Now create Massey ratings for the five team example on page 5. Each team played every other team exactly once, so the off-diagonal elements of $\mathbf{M}$ are -1 and the diagonal elements are 4. But to insure that $\mathbf{M}$ has full rank (and hence a unique least squares solution), use the trick suggested by Massey in [52], and replace the last row with a constraint that forces all ranks to sum to 0. Thus, the Massey least squares system $\bar{\mathbf{M}}\mathbf{r} = \bar{\mathbf{p}}$ is

$$
\begin{pmatrix}
4 & -1 & -1 & -1 & -1 \\
-1 & 4 & -1 & -1 & -1 \\
-1 & -1 & 4 & -1 & -1 \\
-1 & -1 & -1 & 4 & -1 \\
1 & 1 & 1 & 1 & 1
\end{pmatrix}
\begin{pmatrix}
r_1 \\ r_2 \\ r_3 \\ r_4 \\ r_5
\end{pmatrix}
=
\begin{pmatrix}
-124 \\ 91 \\ -40 \\ -17 \\ 0
\end{pmatrix},
$$

which produces the Massey rating and ranking lists given in the following table.

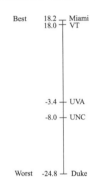

Team	Rating r	Rank
Duke	-24.8	5th
Miami	18.2	1st
UNC	-8.0	4th
UVA	-3.4	3rd
VT	18.0	2nd

Tabular presentation of rating and ranking values is useful, but sometimes the number line representation that is shown on the right-hand side in the figure above is better. The line representation makes it easier to see the relative differences between teams, and this additional information is something we exploit in the section on rating aggregation on 176.

Advanced Features of the Massey Rating Method

These results are reasonable for this tiny dataset, yet there is more to the Massey method [52]. In fact, two vectors more. Massey creates two new vectors, an offensive rating vector $\mathbf{o}$ and a defensive rating vector $\mathbf{d}$, from the overall rating vector $\mathbf{r}$. Massey assumes that team i's overall rating is the sum of its offensive and defensive ratings, i.e., $r_i = o_i + d_i$. Massey squeezes $\mathbf{o}$ and $\mathbf{d}$ from $\mathbf{r}$ using some clever algebra. However, in order to understand the algebra, we need some additional notation. The right-hand side vector $\mathbf{p}$, whose i^{th} element p_i holds team i's cumulative point differentials summed over all games played that season, can be decomposed into $\mathbf{p} = \mathbf{f} - \mathbf{a}$. The vector $\mathbf{f}$ is the "points for" vector, which holds the total number of points scored by each team over the course of the season. The vector $\mathbf{a}$ is called the "points against" vector and it holds the total number of points scored against each team over the course of the season. One additional decomposition is required to understand Massey's method for finding $\mathbf{o}$ and $\mathbf{d}$. The Massey coefficient matrix $\mathbf{M}$ can be decomposed into $\mathbf{M} = \mathbf{T} - \mathbf{P}$, where $\mathbf{T}$ is a diagonal matrix holding the total number of games played by each team and $\mathbf{P}$ is an off-diagonal matrix holding the number of pair-wise matchups between teams during the season. We begin with the original Massey least squares system $\mathbf{Mr} = \mathbf{p}$ and use a string of substitutions to show how the two new rating vectors $\mathbf{o}$ and $\mathbf{d}$ are derived.

$$\mathbf{Mr} = \mathbf{p}$$
$$(\mathbf{T} - \mathbf{P})\mathbf{r} = \mathbf{p}$$
$$(\mathbf{T} - \mathbf{P})(\mathbf{o} + \mathbf{d}) = \mathbf{p}$$
$$\mathbf{To} - \mathbf{Po} + \mathbf{Td} - \mathbf{Pd} = \mathbf{p}$$
$$\mathbf{To} - \mathbf{Po} + \mathbf{Td} - \mathbf{Pd} = \mathbf{f} - \mathbf{a}.$$

Now the final equation above can be split into two separate equations.

$$\mathbf{To} - \mathbf{Pd} = \mathbf{f} \quad \text{and} \quad \mathbf{Po} - \mathbf{Td} = \mathbf{a}.$$

The equation on the left, $\mathbf{To} - \mathbf{Pd} = \mathbf{f}$, says that the total number of points scored by a team over the course of a season can be built by multiplying the offensive score of that team by the number of games played, then subtracting the sum of the defensive scores of its opponents. Working with the equation on the left brings us closer to solving for the two new rating vectors $\mathbf{o}$ and $\mathbf{d}$.

$$\mathbf{To} - \mathbf{Pd} = \mathbf{f}$$
$$\mathbf{T}(\mathbf{r} - \mathbf{d}) - \mathbf{Pd} = \mathbf{f}$$
$$(\mathbf{T} + \mathbf{P})\mathbf{d} = \mathbf{Tr} - \mathbf{f}.$$

Notice that the right-hand side $\mathbf{Tr} - \mathbf{f}$ from the last line above is a vector of constants as $\mathbf{r}$ has already been computed. Thus, the Massey linear system for finding $\mathbf{d}$ (given $\mathbf{r}$) is

$$(\mathbf{T} + \mathbf{P})\mathbf{d} = \mathbf{Tr} - \mathbf{f}.$$

And finally, once both $\mathbf{r}$ and $\mathbf{d}$ are available, $\mathbf{o}$ can be computed using the fact that $\mathbf{r} = \mathbf{o} + \mathbf{d}$.

The Running Example: Advanced Massey Rating Method

Massey's creation and use of $\mathbf{o}$ and $\mathbf{d}$ opens an interesting avenue of research for sports ranking models. Namely, the prediction of point outcomes or at least point spreads. Thus, we apply Massey's method to our small example. In particular, Massey uses these two new rating vectors $\mathbf{o}$ and $\mathbf{d}$ to predict point scores of future matchups. He assumes that o_i gives "the number of points team i is expected to score against an average defense." Combining this with the available defensive ratings, Massey makes point outcome predictions. In particular, if teams i and j play, then the predicted number of points team i would score against team j is given by $o_i - d_j$. Similarly, the predicted number of points team j would score against team i is given by $o_j - d_i$. Table 2.1 gives all three rating vectors for the complete Massey method. Notice that the method predicts a VT vs. Duke game outcome of $20.7 - (-26.8) = 47.5$ to $2.0 - (-2.7) = 4.7$, or roughly 48 to 5. This is rather close to the actual game outcome from the 2005 data, as VT beat Duke 45 to 0 that year. Before getting too excited about the gambling potential of the advanced Massey model, note that not all predictions are this good. For example, the model predicts Miami to beat VT by a score of 25 to 24, which is quite far from the actual game outcome of 25 to 7 from the 2005 data. Part of the blame associated with the volatility of the predictions' success lies with the tiny amount of data (just five teams) used. The issue of point spreads is of interest to many sports fans, particularly the betting types, so a complete discussion of point spreads and optimal "spread ratings" is presented in Chapter 9 on page 113.

Table 2.1 Complete list of Massey rating and ranking vectors

Team	$\mathbf{r}$	Rank with $\mathbf{r}$	$\mathbf{o}$	Offensive Rank	$\mathbf{d}$	Defensive Rank
Duke	-24.8	5th	2.0	4th	-26.8	5th
Miami	18.2	1st	22.0	1st	-3.8	2nd
UNC	-8.0	4th	1.4	5th	-9.4	3rd
UVA	-3.4	3rd	7.8	3rd	-11.2	4th
VT	18.0	2nd	20.7	2nd	-2.7	1st

Summary of the Massey Rating Method

Shown below are the labeling conventions that have been adopted to describe the Massey method for ranking sports teams.

n number of teams in the league = order of the matrices $\mathbf{T}, \mathbf{P}$, and $\mathbf{M}$

m number of games played

$\mathbf{X}_{m \times n}$ game-by-team matrix;
$\quad \mathbf{X}_{ki} = 1$, if team i won in game k, -1, if team i lost, 0, otherwise.

$\mathbf{T}_{n \times n}$ diagonal matrix containing information about the *total* games played;
$\quad \mathbf{T}_{ii} = $ total number of games played by team i during the season

$\mathbf{P}_{n \times n}$ off-diagonal matrix containing *pair-wise* matchup information;
$\quad \mathbf{P}_{ij} = $ number of pair-wise matchups between teams i and j

$\mathbf{M}_{n \times n}$ square symmetric M-matrix called the *Massey Matrix*;
$\quad \mathbf{M} = \mathbf{T} - \mathbf{P}$

$\bar{\mathbf{M}}_{n \times n}$ *adjusted Massey Matrix* that is created when one row of $\mathbf{M}$ is replaced with $\mathbf{e}^T$ (the vector of all ones)

$\mathbf{f}_{n \times 1}$ cumulative *points for* vector;
$\quad f_i = $ total points scored by team i during the season

$\mathbf{a}_{n \times 1}$ cumulative *points against* vector;
$\quad a_i = $ total points scored against team i during the season

$\mathbf{p}_{n \times 1}$ cumulative *point differential* vector;
$\quad p_i = f_i - a_i$; cumulative point differential of team i during the season

$\bar{\mathbf{p}}_{n \times 1}$ *adjusted point differential* vector created when one element of $\mathbf{p}$ is replaced with a 0

$\mathbf{r}_{n \times 1}$ general rating vector produced by the Massey least squares system

$\mathbf{o}_{n \times 1}$ offensive rating vector produced by the Massey least squares system

$\mathbf{d}_{n \times 1}$ defensive rating vector produced by the Massey least squares system

> **Massey Least Squares Algorithm**
>
> 1. Solve the system $\bar{\mathbf{M}}\mathbf{r} = \bar{\mathbf{p}}$ to obtain the general Massey rating vector $\mathbf{r}$.
> 2. Solve the system $(\mathbf{T} + \mathbf{P})\mathbf{d} = \mathbf{Tr} - \mathbf{f}$ for the defensive rating vector $\mathbf{d}$.
> 3. Compute the offensive rating vector $\mathbf{o} = \mathbf{r} - \mathbf{d}$.

Listed below are some properties of the Massey method.

- The Massey method uses point score data to rate teams. However, other game statistics could be used. And in other contexts, any data on pair-wise comparisons of items could be used. For example, to rate the trading power of nations, the gross number of exports could be used. The nation with the larger number of exports (perhaps normalized by accounting for population) would win in the hypothetical matchup. In this way, the United Nations' HDI measure of page 6 could be augmented and enhanced.

- The Massey method creates three vectors, the general rating vector $\mathbf{r}$, the offensive rating vector $\mathbf{o}$, and the defensive rating vector $\mathbf{d}$. By cleverly combining $\mathbf{o}$ and $\mathbf{d}$, one can estimate point spreads. (Also see Chapter 9 on page 113.)

ASIDE: The Massey Method for Ranking Webpages

In this aside we apply the Massey method to webpages to emphasize the point that this method can be used to rank any collection of items. (Actually, with just a little thought, the same is true of every method in this book.) Because webpages are presented in response to user queries based on their ranked order, webpage rankings are a fundamental part of every major search engine. For instance, Figure 2.1 shows the webpages ranked in positions 1 through 10 for a query to Google on the phrase "Massey ranking."

The trick to applying the Massey method to the Web is to define the notion of a game or pair-wise matchup for webpages. There are numerous possibilities. For example, if statistics on webpage traffic are available with the Alexa search engine [78, 49], then we can say that webpage i beats webpage j if $t_i > t_j$, where t_i and t_j are the traffic measures for these pages. In this case, $t_i - t_j$ represents the "point" differential. The famous PageRank [15] measure π_i for webpage i is another measure that can be used to derive matchup information. In this case, webpage i beats webpage j by $\pi_i - \pi_j$ points if $\pi_i > \pi_j$. The Massey web ranking method begins with the same fundamental idealized equation

$$r_i - r_j = y_k,$$

except the margin of victory y_k for game k is modified depending on the data used. For example, $y_k = t_i - t_j$ if traffic measures are used and $y_k = \pi_i - \pi_j$ if PageRank measures are used to determine winners of hypothetical matchups.

The Massey web ranking method then proceeds as usual. And since the Massey method can be used to create two rating vectors, the offensive and the defensive vectors, each webpage

Google

Search   About 368,000 results (0.26 seconds)

Everything

Images

Maps

Videos

News

Shopping

More

Princeton, NJ
Change location

Show search tools

Massey Ratings
masseyratings.com/
Computer ratings for virtually every sport: college football, basketball, hockey, and
baseball, NFL, NBA, NHL, MLS, WNBA, and more. Links to other **rankings**, ...

College Football Ranking ...
Pigskin GB Eldredge Bassett Mark
Trombetta MARS ...

Massey Ratings - NFL
Least Likely Results. 1 9/11 Buffalo
(10) 41 at Kansas City ...

College Basketball Ranking ...
Pomeroy USA Today Coaches Dokter
Entropy DeSimone ...

MLB
Least Likely Results. 1 6/30 Astros
(30) 7 Rangers (4) 0 ...

Massey Ratings - CF
Click here or look at the -BCS-
column to see the **ranking** ...

FAQ
the "**Massey** Ratings", which utilize
actual game scores and ...

More results from masseyratings.com »

MASSEY RATINGS - HighSchoolSports.net
www.highschoolsports.net/**massey**/
Oct 15, 2009 – High school Football scores, **rankings**, schedules and players for every
... Ken **Massey's** renowned ratings algorithm was added to the BCS ...

Massey-Peabody Power **Rankings**
www.**massey**-peabody.com/
NFL Power **Rankings** Rufus Peabody Cade **Massey Massey**-Peabody Sabremetrics
Judgment under uncertainty Forecasting Outcome bias Overconfidence ...

Massey University | **Ranking** & Review
www.4icu.org › Oceania › New Zealand
Sep 5, 2011 – 10200 Universities > New Zealand > **Massey** University web **ranking** &
review including accreditation, courses, tuition, admission, size, facilities ...

The **Massey**-Peabody Power **Rankings** - WSJ.com
online.wsj.com/.../SB10001424052970204485304576643362701890...
Oct 21, 2011 – Cade **Massey**, an assistant professor at the Yale School of Management,
and Rufus Peabody, a Las Vegas sports analyst, developed this ...

Massey University - Topuniversities
www.topuniversities.com/institution/**massey**-university
Since 2004 our **rankings** have helped students understand which universities ... **Massey**
University is one of New Zealand?s leading educational institutions. ...

Massey University New Zealand **Ranking** and Reviews
www.hotcoursesabroad.com/study/.../**massey**.../international.html
International students: Find out about **Massey** University New Zealand Palmerston
North. Read **Massey** University New Zealand Palmerston North reviews, ...

Massey-Peabody Week 5 NFL Power **Rankings** | SNY Why Guys
www.snywhyguys.com/.../**massey**-peabody-week-5-nfl-power-rankin...
Oct 8, 2011 – I urge you all to visit the **Massey**-Peabody website. But I didn't want these
to go to waste because they are the best power **rankings** in the ...

Kenneth **Massey's** Initial 2011 Preseason **Ranking** Comparison
www.blog.rogerspoll.com/.../kenneth-**masseys**-initial-2011-preseason...
Aug 1, 2011 – Kenneth **Massey** has released his initial 2011 **ranking** comparison,
building a consensus from 15 different **rankings** made available on or ...

Figure 2.1 Results for query of "massey ranking" to Google search engine

will have dual ratings. These dual ratings echo of hubs and authorities, the dual rankings that are so well-known within the web ranking community [45, 49]. Chapter 7 talks more about hubs and authorities in the sports ranking context.

Chapter 12 presents a time-weighting refinement to the Massey model that also has exciting potential for the Web Massey method. This refinement provides a natural way to incorporate time into the ranking. With respect to the Web, this means that available time data, such as time stamps marking the most recent update to a webpage, can be used.

Spam is an increasing concern for many of today's search engines—yet a ranking method such as the Massey method is much less susceptible to spam than many of the current webpage ranking methods since it can be built from measures such as site traffic over which webpage owners have less control. This is contrasted with the link-based measures used by nearly all major search engine rankings today. Webpage owners certainly control their own outlinks, and have learned some clever tricks to influence their inlinks too [49]. We believe that the best most spam-resistant rankings can be built using a combination of webpage measures as detailed by the aggregation methods of Chapters 6 and 14.

Thus far in this aside we have painted a very promising picture of the applicability of the Massey method for ranking webpages. Now we turn to one final issue, the very practical issue of scale, an issue that can make or break a method. A fantastic method that does not scale up to web-sized graphs with millions and billions of nodes is useless for ranking webpages. Fortunately, the Massey method contains some exploitable structure that makes it web-able. We describe one possible scenario for building the Massey linear system directly from the hyperlink structure of the Web. This hyperlink data is readily available as it is the basic data used by nearly all current search engines. A more sophisticated model could use more sophisticated data such as the traffic and PageRank measures mentioned above. For now, we use the basic hyperlink data to demonstrate the scalability of the Massey method to web proportions.

Let $\mathbf{W}$ be a weighted adjacency matrix for the Web. That is, w_{ij} is the weight of the directed hyperlink between webpages i and j; Thus, $w_{ij} = 0$ if there is no hyperlink between webpages i and j. In this implementation of the Web Massey method, $\mathbf{W}$ holds the "point" differential information so that $w_{ij} > 0$ means webpage i outlinks to webpage j. The Massey matrix $\mathbf{M}$ has dimension $n \times n$ where n is the number of webpages. Since m_{ij} is the negation of the number of times webpages i and j match up, $m_{ij} = m_{ji} = -1$ whenever $w_{ij} > 0$. Thus, if the weighted adjacency matrix is available, the Web Massey system can be built with just a few additional lines of code, which we write in MATLAB style.

```
A = W' > 0;    % A is a binary adjacency matrix built from the
                  transpose of the weighted adjacency matrix W
M = -A - A' + diag(Ae) + diag(A'e);  % Web Massey matrix M
```

The final piece of the Massey linear system $\mathbf{Mr} = \mathbf{p}$ is the right-hand side vector $\mathbf{p}$. Like $\mathbf{M}$, $\mathbf{p}$ can be built easily. The ith element of $\mathbf{p}$, denoted p_i, is the sum of all inlinks (or alternatively the sum of weights of all inlinks) into page i minus the sum of all outlinks from page i. Thus, for the Massey Web system

```
p = W'e - We.
```

While the linear system $\bar{\mathbf{M}}\mathbf{r} = \bar{\mathbf{p}}$ that must be solved is enormous, it is no more so than current ranking systems such as PageRank or HITS [49]. In fact, it can be stored and solved efficiently with an iterative method such as Jacobi, SOR, or GMRES [32, 75, 74, 66].

ASIDE: BCS Ratings

The Bowl Championship Series (or BCS) ratings determine who gets to play for the big bucks in the post-season college football bowl games. So how are these important ratings determined? They are composed of three equally weighted components. Two of the components are human polls (the Harris Interactive College Football Poll and the *USA Today* Coaches Poll), and the third is an aggregation of six computer rankings provided by Jeff Anderson and Chris Hester, Richard Billingsley, Wesley Colley, Kenneth Massey, Jeff Sagarin, and Peter Wolfe.

A voting technique known as *Borda counts* is used to merge these rankings—a complete description of Borda counts is given on page 165. For the BCS it works like this. In the human polls (conducted weekly between October and December) the voting members fill out a top 25 rating ballot. Each team is given a rating $1 \leq r \leq 25$ that corresponds in reverse order to its rank—i.e., on each ballot the first place team receives a rating of $r = 25$, the second place team is given a rating of $r = 24$, etc., until the last place team gets a rating of $r = 1$. The same scoring system is applied to the computer rankings. In each human poll, teams are evaluated on their total rating—i.e., the sum of their ratings across all ballots. The number of voters, which can vary, is accounted for by computing each team's "BCS quotient" that is defined to be its percentage of a perfect score. This is made clear in the following example.

- If the Harris Interactive College Football Poll has 113 voting members for a given season, then the highest total rating that a team can achieve is $113 \times 25 = 2,825$, which is attained if and only if all 113 members give that team a rating of $r = 25$, or equivalently, they all rank that team as #1. Similarly, the lowest total rating that a team can have is 113, obtained when each member assigns a rating of $r = 1$ to that team (i.e., everyone ranks the team #25). So, if r_i is the Harris rating given to a team (say NC State) by the ith voter, then

$$\text{The Harris BCS quotient for NC State} = \frac{1}{2825} \sum_{i=1}^{113} r_i.$$

- If the *USA Today* Coaches Poll has 59 voting members for the same season, then the highest possible total rating is $59 \times 25 = 1,475$. Consequently, if r_i is the *USA Today* Coaches rating given to NC State by the ith voter, then

$$\text{The } USA\ Today \text{ Coaches BCS quotient for NC State} = \frac{1}{1475} \sum_{i=1}^{59} r_i.$$

- Each of the six computer models provides numerical ratings for their top 25 teams just as the humans do—i.e., $r = 25$ for the highest ranked team, $r = 24$ for the second ranked team, etc. However, before computing the total sum of ratings for each team, the highest and lowest ratings for each team are dropped. Consequently, each team has only four computer ratings. These four ratings are summed to provide the total rating for each team. Thus the highest total computer rating that can be achieved is $4 \times 25 = 100$. For example, if the ratings for NC State from the six computer models are ordered as $r_1 \geq r_2 \geq r_3 \geq r_4 \geq r_5 \geq r_6$, then

$$\text{The computer BCS quotient for NC State} = \frac{1}{100}(r_2 + r_3 + r_4 + r_5).$$

Putting It All Together. The BCS ratings and rankings at the end of each polling period are determined simply by adding together the BCS quotients defined above. While it makes no difference in the rankings, the average of the three BCS quotients is sometimes published instead of the sum of the BCS quotients because the average forces all final ratings to be in the interval $(0, 1]$. In other words,

$$\text{Final BCS Rating} = \frac{\text{Harris Quotient} + \text{Coaches Quotient} + \text{Computer Quotient}}{3}.$$

The Recent BCS Computer Models. The computer models that have recently been used for the BCS computer rankings include those by the people listed below. However, the specifics of their methods are generally not available. Many developers do not want the details of their secret potions completely revealed, so they only hint at what they do.

1. Jeff Anderson and Chris Hester (`www.andersonsports.com`) give very little in the way of details other than to say that they reward teams for beating quality opponents, that they do not consider the margin of victory, and that their ratings depend not only on win-loss records but also on conference strength.

2. Richard Billingsley (`www.cfrc.com`) divulges nothing. It is theorized by others that the main components in his formula are win-loss records, opponents' strength based on their records, ratings, and ranks (with a strong emphasis on most recent performances), and minor considerations for the site of the game and defensive performance.

3. Wesley Colley's (`www.colleyrankings.com`) techniques are detailed in Chapter 3 on page 21.

4. Kenneth Massey's early techniques are described on page 9, but he suggests that he now uses more refined (but not completely specified) methods—see `masseyratings.com`.

5. Jeff Sagarin (`www.usatoday.com/sports/sagarin.htm`) says that his model primarily utilizes the Elo system that is described in detail in Chapter 5 on page 53.

6. Peter Wolfe (`prwolfe.bol.ucla.edu/cfootball/ratings.htm`) states that his method is adapted from methods of Bradley and Terry in [13]. He says that it is a maximum likelihood scheme in which team i is assigned a rating value π_i that is used in predicting the expected result when team i meets team j, with the likelihood of i beating j given by $\pi_i/(\pi_i + \pi_j)$. The probability P of all the results happening as they actually did is the product of multiplying together all the individual probabilities derived from each game. The rating values are chosen in such a way that the value of P is maximized. A similar technique is discussed by James Keener in [42].

ASIDE: No Wiggle Room Allowed

The BCS ratings narrow the competitors in the NCAA Division I Football Bowl Subdivision (or FBS—formerly Division I-A) to two teams that compete in the "BCS National Championship Game." Voting members from the American Football Coaches Association are contractually obligated to vote the winner of this game as the "BCS National Champion." Moreover, a contract signed by each conference requires them to recognize the winner of this game as the "official and only champion," thus eliminating the possibility of multiple champions being crowned by the competing rating entities.

ASIDE: The Notre Dame Rule

The BCS ratings are used to determine which schools are given berths in post-season bowl games, but the ratings are tempered by a complicated set of tangled rules designed to protect various athletic conferences and historically powerful programs. Perhaps the most bizarre example is the "Notre Dame rule" that gives Notre Dame an automatic berth in the bowl games if it finishes in the top eight. And it get stranger. Notre Dame is also guaranteed a separate and unique payment reported to be 1/66th of the net BCS bowl revenues—regardless of their win-loss record for the season and regardless of whether or not they qualify for a bowl game! It was reported that the Fighting Irish were paid approximately $1,700,000 in 2009, a season in which they had only a 6-6 record and fired their coach, Charlie Weis. It has been estimated that if they had made a bowl game, then they would have been paid about $6,000,000.

The Notre Dame rule is an ongoing controversy. The exclusion of almost half the teams in the FBS from any realistic opportunities to win or compete for the BCS Championship while giving Notre Dame unique advantages and opportunities sticks in the craw of many fans.

ASIDE: Obama and the Department Of Justice

During his 2008 presidential campaign, then President-elect Obama appeared on Monday Night Football. ESPN sportscaster Chris Berman asked him to describe the one thing about sports that he would change, and Obama mentioned his disgust with using the BCS rating system to determine bowl-game participants. Instead, like many fans, he favored an eight-team playoff to determine the winner of the national title. Many fans wish the president would simply sign this suggestion into law.

In 2009 Senator Orrin Hatch of Utah asked the U.S. Justice Department to review the legality of the BCS for possible violations to antitrust laws. Hatch was motivated by the fact that the University of Utah had been undefeated but was denied a chance to play in the 2008 title game due to the peculiarities of the BCS system. Hatch and others cite antitrust issues that are at the heart of the "tremendous inequities," citing the BCS's favoritism of some conferences. For instance, under BCS rules, the champions of six conferences have automatic bids to play in the top-tier bowl games.

Atlantic Coast Conference	⟶	Orange Bowl
Big 12 Conference	⟶	Fiesta Bowl
Big East Conference	⟶	No specific BCS Bowl
Big Ten Conference	⟶	Rose Bowl
Pacific-12 Conference	⟶	Rose Bowl
Southeastern Conference	⟶	Sugar Bowl

In addition, these six conferences receive more money for these appearances than other conferences. However, conference alignments are currently in a state of flux, so several changes may have occurred by the time you read this.

As of May 2011, the BCS was put on notice that they are being scrutinized by the U. S. Department of Justice. Assistant Attorney General Christine Varney, the department's antitrust division leader, reportedly sent a two-page letter to Mark Emmert (NCAA president at the time) stating that, "Serious questions continue to arise suggesting that the current Bowl Championship Series (BCS) system may not be conducted consistent with the competition principles expressed in federal antitrust laws."

By The Numbers —

$21,200,000 = amount received by each BCS conference in 2010.

$6,000,000 = additional amount paid to the SEC and Big Ten
(they each had two teams in a BCS game).

— illegalshift.com

Chapter Three

Colley's Method

In 2001, Dr. Wesley Colley, an astrophysicist by training, wrote a paper about his new method for ranking sports teams [22]. This side project of his became so successful that, like one of Massey's models, it too is now incorporated in the BCS method of ranking NCAA college football teams. His method, which we call the Colley Rating Method, is a modification of one of the simplest and oldest rating systems, the rating system that uses winning percentage. Winning percentage rates team i with the value r_i according to the rule

Wesley Colley

$$r_i = \frac{w_i}{t_i},$$

where w_i and t_i are the number of wins and total number of games played by team i, respectively. The winning percentage method is the prevailing rating method for recreational leagues and tournaments worldwide. In fact, it is also used by most professional leagues. Though simple and easy to use, this rating system does have a few obvious flaws. First, ties in the ratings often occur in sports such as football where most teams play the same number of games against the same set of teams. Second, the strength of opponents is not factored into the analysis in any way. Defeating the weakest opponent in the league earns a team the same advance in the ratings that defeating the strongest team does. This is arguably unfair. Third, there are times when winning percentage ratings give unusual results. For instance, at the beginning of the season all teams have preseason ratings of $\frac{0}{0}$ and, in addition, as the season progresses a winless team has a rating of 0. To possibly remedy some of these flaws, Wesley Colley proposed a method, which we call the Colley method, for ranking sports teams.

The Main Idea behind the Colley Method

The Colley method begins with a slight modification to the traditional winning percentage formula so that

$$r_i = \frac{1 + w_i}{2 + t_i}. \tag{3.1}$$

In this section we show that the main advantage of this modification is the consideration of strength of schedule, i.e., the strength of a team's opponents.

Colley's adjustment to the winning percentage formula comes from Laplace's "rule of succession" [29] used to find a marker on a craps table. Though this modification to the traditional winning percentage formula appears quite modest, it has several advantages over the traditional formula. Instead of the nonsensical preseason ratings of $\frac{0}{0}$, each team now begins the season with equal ratings of $\frac{1}{2}$. Further, when team i loses to an opponent in week one of the season, Colley's rating produces $r_i = \frac{1}{3}$, which he argues is more reasonable than $r_i = 0$.

The next advantage to using Laplace's rule (3.1) rather than standard winning percentage is tied to the notion of strength of schedule. The idea here is that a team ought to receive a greater reward for beating strong opponents as opposed to weak opponents. In effect, the rating of team i should be connected to the ratings of its opponents. Colley argues that Laplace's rule contains, though hidden away, the strength of team i's schedule. Notice in Equation (3.1) that all teams start with $r_i = 1/2$, and as the season progresses the ratings deviate above or below this starting point. In fact, one team's gain (in the form of a win) is another's loss. As a result these ratings are interdependent. Such interdependence is not apparent from (3.1) and can only be uncovered with a careful dissection. First we decompose the number of games won by a team.

$$
\begin{aligned}
w_i &= \frac{w_i - l_i}{2} + \frac{w_i + l_i}{2} \\
&= \frac{w_i - l_i}{2} + \frac{t_i}{2} \\
&= \frac{w_i - l_i}{2} + \sum_{j=1}^{t_i} \frac{1}{2}.
\end{aligned}
$$

Because all teams start with $r_j = 1/2$ at the beginning of the season, the summation $\sum_{j=1}^{t_i} \frac{1}{2}$ is initially equal to $\sum_{j \in O_i} r_j$, where O_i is the set of opponents for team i. As the season progresses, the summation $\sum_{j=1}^{t_i} \frac{1}{2}$ is not exactly equal to $\sum_{j \in O_i} r_j$ but can be well-approximated by the cumulative ratings of a team's opponents. (Recall that the ratings always hover about 1/2.) As a result,

$$
w_i \approx \frac{w_i - l_i}{2} + \sum_{j \in O_i} r_j.
$$

Assuming equality and inserting this into (3.1) produces

$$
r_i = \frac{1 + (w_i - l_i)/2 + \sum_{j \in O_i} r_j}{2 + t_i}. \tag{3.2}
$$

Of course, as we have seen with the rating systems in previous chapters, the goal is to find the unknown r_i's. Equation (3.2) shows that, in this case, the unknown r_i depends on other unknowns, the r_j's. And this also reveals how Colley's method incorporates strength of opponents into a team's ratings. With a little more algebra and the help of matrix notation, we find that the dependency of r_i on the other r_j's is not a problem. The r_i's can be computed easily. In fact, Colley Equation (3.2) can be written compactly as a linear system $\mathbf{Cr} = \mathbf{b}$, where $\mathbf{r}_{n \times 1}$ is the unknown Colley rating vector, $\mathbf{b}_{n \times 1}$ is the right-hand side vector defined as $b_i = 1 + \frac{1}{2}(w_i - l_i)$ and $\mathbf{C}_{n \times n}$ is the Colley coefficient matrix defined as

$$\mathbf{C}_{ij} = \begin{cases} 2 + t_i & i = j, \\ -n_{ij} & i \neq j, \end{cases}$$

where n_{ij} is the number of times teams i and j played each other. It can be proven that the Colley system $\mathbf{Cr} = \mathbf{b}$ always has a unique solution since $\mathbf{C}_{n \times n}$ is invertible.

The Running Example

It is time to test the Colley method on our small example using the data from Table 1.1. The Colley linear system $\mathbf{Cr} = \mathbf{b}$ is below.

$$\begin{pmatrix} 6 & -1 & -1 & -1 & -1 \\ -1 & 6 & -1 & -1 & -1 \\ -1 & -1 & 6 & -1 & -1 \\ -1 & -1 & -1 & 6 & -1 \\ -1 & -1 & -1 & -1 & 6 \end{pmatrix} \begin{pmatrix} r_1 \\ r_2 \\ r_3 \\ r_4 \\ r_5 \end{pmatrix} = \begin{pmatrix} -1 \\ 3 \\ 1 \\ 0 \\ 2 \end{pmatrix}.$$

Note that $\mathbf{C}$ is a real symmetric positive definite matrix. These properties mean that $\mathbf{C}$ has a Cholesky decomposition [54] such that $\mathbf{C} = \mathbf{U}^T\mathbf{U}$, where $\mathbf{U}$ is an upper triangular matrix. As a result, if the Cholesky factorization is available, the Colley system $\mathbf{Cr} = \mathbf{b}$ can be solved especially efficiently. However, for many sports applications, the systems are small enough to allow software packages such as MATLAB to quickly compute the rating vector $\mathbf{r}$ by standard numerical routines, e.g., Gaussian elimination, Krylov methods, etc. Applying the Colley rating method to our running example produces the ratings, and hence, rankings, displayed in Table 3.1.

Table 3.1 Colley rating results for the 5-team example

Team	r	Rank
Duke	.21	5th
Miami	.79	1st
UNC	.50	3rd
UVA	.36	4th
VT	.65	2nd

The most glaring attribute of the Colley method is that game scores are not considered in any way. Depending on your point-of-view, this is either a strength or a weakness of the model. Note that UNC and UVA are swapped when compared with the Massey rankings from page 13. This is an interesting consequence of avoiding game scores. The Massey method favored UVA over UNC due to the superior overall point differential of UVA, despite UNC's better record. Colley argues that by ignoring game scores, his method is *bias-free*, a word he uses to refer to the method's avoidance of the potential rating problem created when strong teams run up the score against weak teams. Methods, such as the Massey method, that use game scores can be subject to such bias.

Summary of the Colley Rating Method

$\mathbf{C}_{n \times n}$ real symmetric positive definite matrix called the *Colley matrix*;

$$\mathbf{C}_{ij} = \begin{cases} 2 + t_i & i = j, \\ -n_{ij} & i \neq j. \end{cases}$$

t_i total number of games played by team i

n_{ij} number of times teams i and j faced each other

$\mathbf{b}_{n \times 1}$ right-hand side vector; $b_i = 1 + \frac{1}{2}(w_i - l_i)$

w_i total number of wins accumulated by team i

l_i total number of losses accumulated by team i

$\mathbf{r}_{n \times 1}$ general rating vector produced by the Colley system

n number of teams in league = order of $\mathbf{C}$

Colley's Algorithm

Solve the system $\mathbf{Cr} = \mathbf{b}$ to obtain the Colley rating vector $\mathbf{r}$.

Listed below are some properties of the Colley method.

- The results of the Colley method are *bias-free*, meaning that they are generated using only win-loss information and not point score data. Thus, the Colley ratings are unaffected by the existence of teams that purposefully run up the score against weak opponents. Furthermore, some sports or leagues are more prone to large point differentials than others. For example, the differential in a typical NBA game is generally smaller than that in an NCAA basketball game. Consequently, the suitability of the Colley method may depend on the league, sport, or application in question.

- The Colley ratings follow a conservation property. Each team begins the season with an initial rating of 1/2, and as the season progresses a team bounces back and forth above and below this center point depending on its game outcomes. Yet overall, the average $\bar{r}$ of all ratings r_i (i.e., $\bar{r} = \mathbf{e}^T \mathbf{r}/n$) remains at 1/2. Thus, there is an overall conservation of the total rating. When one team's rating improves, another team's must suffer.

- The Colley method is also well-suited to non-sports applications in which, in an analogous fashion, the equivalent of point differential data is unavailable or undesirable.

Connection between Massey and Colley Methods

While seemingly quite different in philosophy, there is a striking connection between the Massey and the Colley methods. The methods are related by the formula $\mathbf{C} = 2\mathbf{I} + \mathbf{M}$. As a result, it is easy to Colleyize the Massey method or vice versa. For example, the original Massey method of $\mathbf{Mr} = \mathbf{p}$ can be Colleyized as

$$(2\mathbf{I} + \mathbf{M})\mathbf{r} = \mathbf{p} \quad \text{(which is also } \mathbf{Cr} = \mathbf{p}.\text{)}$$

The Colleyized Massey method uses a right-hand side of $\mathbf{p}$, which contains point score information rather than a right-hand side of $\mathbf{b}$, which uses only win-loss information. Further, the addition of $2\mathbf{I}$ to the coefficient matrix adds the Laplace trick and also makes the system nonsingular so that Massey's "replace an equation" method for removing the singularity is not necessary. The Colleyized Massey method $\mathbf{Cr} = \mathbf{p}$ is no longer bias-free because point scores have now been incorporated by the use of $\mathbf{p}$. The Colleyized Massey method applied to the running example produces the rating vector $\mathbf{r}$ shown in Table 3.2, which produces the same rank ordering as the standard Massey method with numerical values that are slightly different from those in the table on page 11.

Table 3.2 Colleyized Massey rating results for the 5-team example

Team	r	Rank
Duke	-17.7	5th
Miami	13.0	1st
UNC	-5.7	4th
UVA	-2.4	3rd
VT	12.9	2nd

In a similar fashion, the Colley method can be Masseyized so that the Masseyized Colley method solves the linear system $\mathbf{Mr} = \mathbf{b}$.

ASIDE: Movie Rankings: Colley and Massey Meet Netflix

Netflix is a movie rental company that operates online through its virtual location, www.netflix.com. For a monthly subscription fee, Netflix members can choose a certain number of movies to rent each month. These are mailed directly to users and, once viewed, are returned by mail to Netflix. Periodically, Netflix asks users to rate the movies they have rented. This user-movie information is collected and analyzed to help the company make future recommendations to their users. In 2007, Netflix offered a $1 million prize to the individual or group that could improve their current recommendation system by 10%. A small sample of the data that was released for the competition appears below.

$$\mathbf{U} = \begin{array}{c} \text{User 1} \\ \text{User 2} \\ \text{User 3} \\ \text{User 4} \\ \text{User 5} \\ \text{User 6} \end{array} \begin{pmatrix} 5 & 4 & 3 & 0 \\ 5 & 5 & 3 & 1 \\ 0 & 0 & 0 & 5 \\ 0 & 0 & 2 & 0 \\ 4 & 0 & 0 & 3 \\ 1 & 0 & 0 & 4 \end{pmatrix}.$$

Movie 1 Movie 2 Movie 3 Movie 4

$\mathbf{U}$ is a user-by-movie matrix containing ratings. Valid ratings are integers 1 through 5, with 5 being the best score. A 0 means the user did not rate the movie. To apply the ideas of sports ranking to such a matrix we must think in terms of pair-wise matchups between movies. For instance, in order to apply the Colley method to rank movies, we create the movie-movie

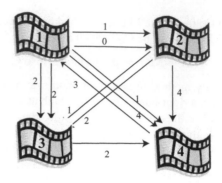

Figure 3.1 Movie-movie graph associated with the Massey method

graph of Figure 3.1 from the rating matrix **U**. Notice that there are two links between movies 1 and 2. One link has weight 1, because user 1 rated movie 1 one point higher than movie 2. The other link has weight 0 because user 2 also rated both movies, but with the same value. The graph of Figure 3.1 has a link between two movies each time a user rated both movies, one at the head of the link and the tail of the link.

As stated on page 24, the Colley system $\mathbf{Cr} = \mathbf{b}$ can be built with the Massey matrix by using the identity $\mathbf{C} = 2\mathbf{I} + \mathbf{M}$. The Massey matrix associated with Figure 3.1 is

$$
\mathbf{M} = \begin{array}{c} 1 \\ 2 \\ 3 \\ 4 \end{array}\begin{pmatrix} \begin{array}{cccc} 1 & 2 & 3 & 4 \end{array} \\ 7 & -2 & -2 & -3 \\ -2 & 5 & -2 & -1 \\ -2 & -2 & 5 & -1 \\ -3 & -1 & -1 & 5 \end{pmatrix} \implies \mathbf{C} = 2\mathbf{I} + \mathbf{M} = \begin{array}{c} 1 \\ 2 \\ 3 \\ 4 \end{array}\begin{pmatrix} \begin{array}{cccc} 1 & 2 & 3 & 4 \end{array} \\ 9 & -2 & -2 & -3 \\ -2 & 7 & -2 & -1 \\ -2 & -2 & 7 & -1 \\ -3 & -1 & -1 & 7 \end{pmatrix}.
$$

The Colley and Massey right-hand side vectors $\mathbf{b}$ and $\mathbf{p}$, respectively, are

$$
\mathbf{b} = \begin{pmatrix} 3 \\ 2 \\ -.5 \\ -.5 \end{pmatrix} \quad \text{and} \quad \mathbf{p} = \begin{pmatrix} 7 \\ 6 \\ -5 \\ -8 \end{pmatrix}.
$$

Table 3.3 shows the Colley and Massey ratings of these four movies.

Table 3.3 Colley and Massey rating and ranking vectors for four-movie example

Movie	Colley **r**	Colley rank	Massey **r**	Massey rank
1	.67	1st	0.65	2nd
2	.63	2nd	1.01	1st
3	.34	4th	-0.55	3rd
4	.35	3rd	-1.11	4th

Ties are an issue in this Netflix dataset. Because valid movie ratings are integers from 1 to 5, it will often happen that two users rated the two movies with the same score. If we think such matchups tell us very little about which movie dominated the other, we can simply remove ties from the dataset. In this case, the "no ties" $\mathbf{M}$ and $\mathbf{C}$ matrices are below.

$$\mathbf{M} = \begin{array}{c} \\ 1 \\ 2 \\ 3 \\ 4 \end{array} \begin{array}{cccc} 1 & 2 & 3 & 4 \\ \left(\begin{array}{cccc} 6 & -1 & -2 & -3 \\ -1 & 4 & -2 & -1 \\ -2 & -2 & 5 & -1 \\ -3 & -1 & -1 & 5 \end{array} \right) \end{array} \quad \text{and thus,} \quad \mathbf{C} = \begin{array}{c} \\ 1 \\ 2 \\ 3 \\ 4 \end{array} \begin{array}{cccc} 1 & 2 & 3 & 4 \\ \left(\begin{array}{cccc} 8 & -1 & -2 & -3 \\ -1 & 6 & -2 & -1 \\ -2 & -2 & 7 & -1 \\ -3 & -1 & -1 & 7 \end{array} \right) \end{array}.$$

These produce the "no ties" ratings and rankings given in Table 3.4. Notice that the Colley ranking of these four movies changes according to the handling of ties. This makes the point that ties must be handled carefully, an issue that we take up later in Chapter 11.

Table 3.4 Colley and Massey "no ties" rating and ranking vectors for four-movie example

Movie	Colley $\mathbf{r}$	Colley rank	Massey $\mathbf{r}$	Massey rank
1	.28	2nd	0.60	2nd
2	.33	1st	1.09	1st
3	.05	3rd	-0.55	3rd
4	-.39	4th	-1.13	4th

We ran both the Colley and Massey methods on a sample of data from the Netflix Prize competition. This data sample contains all 17,770 movies in the Netflix database (as of 2007), but only a subset of the users. The 13,141 "superusers," users ranking a thousand or more movies, were selected.[1] The fifteen top rated movies according to each method are listed in Table 3.5. Netflix data comes with time stamps marking when a user rated the movie, so improvements are possible by incorporating the time weighting ideas introduced later in Chapter 12.

---

[1] We thank David Gleich for providing us with this dataset.

Table 3.5 Top-25 movies ranked by the Colley and Massey methods

Rank	by the Colley method	by the Massey method
1	*Lord of the Rings III: Return of ...*	*Lord of the Rings III: Return of ...*
2	*Lord of the Rings II: The Two ...*	*Lost*: Season 1
3	*Lost*: Season 1	*Lord of the Rings II: The Two ...*
4	*Star Wars V: Empire Strikes Back*	*Battlestar Galactica*: Season 1
5	*Battlestar Galactica*: Season 1	*Star Wars V: Empire Strikes Back*
6	*Raiders of the Lost Ark*	*Raiders of the Lost Ark*
7	*Star Wars IV: A New Hope*	*Star Wars IV: A New Hope*
8	*The Shawshank Redemption*	*The Shawshank Redemption*
9	*The Godfather*	*The Godfather*
10	*Star Wars VI: Return of the Jedi*	*Star Wars VI: Return of the Jedi*
11	*The Sopranos*: Season 5	*The Sopranos*: Season 5
12	*Schindler's List*	*GoodFellas*
13	*Lord of the Rings I: Fellowship ...*	*Band of Brothers*
14	*Band of Brothers*	*The Simpsons*: Season 5
15	*The Sopranos*: Season 1	*Lord of the Rings I: Fellowship ...*
16	*The Simpsons*: Season 5	*Schindler's List*
17	*Gladiator*	*The Sopranos*: Season 1
18	*CSI*: Season 4	Finding Nemo
19	*The Sopranos*: Season 2	*The Simpsons*: Season 6
20	*The Sopranos*: Season 4	*Gladiator*
21	*The Simpsons*: Season 6	*The Simpsons*: Season 4
22	*The Simpsons*: Season 4	*The Sopranos*: Season 2
23	*Finding Nemo*	*The Godfather*
24	*Toy Story*	*The Sopranos*: Season 4
25	*The Silence of the Lambs*	*CSI*: Season 4

By The Numbers —

$2, 782, 275, 172 =$ highest gross amount ever earned by a movie (as of 2011).
    — *Avatar* (released 2009).

$1, 843, 201, 268 =$ second highest amount (as of 2011).
    — *Titanic* (released 1997).

— en.wikipedia.org

Chapter Four

Keener's Method

James P. Keener proposed his rating method in a 1993 SIAM Review article [42]. Keener's approach, like many others, utilizes nonnegative statistics that result from contests (games) between competitors (teams) to create a numerical rating for each team. In some circles these are called *power ratings*. Of course, once a numerical rating for each team is established, then ranking the teams in order of their ratings is a natural consequence.

Keener's method is to relate the *rating* for a given team to the *absolute strength* of the team, which in turn depends on the *relative strength* of the team—i.e., the strength of the team relative to the strength of the teams that it has played against. This will become more clear as we proceed. Keener bases everything on the following two stipulations that govern the relationship between a team's *strength* and its *rating*.

Strength and Rating Stipulations

1. The *strength* of a team should be gauged by its interactions with opponents together with the strength of these opponents.

2. The *rating* for each team in a given league should be uniformly proportional to the strength of the team. More specifically, if s_i and r_i are the respective strength and rating values for team i, then there should be a proportionality constant λ such that $s_i = \lambda r_i$, and λ must have the same value for each team in the league.

Selecting Strength Attributes

To turn the two rules above into a mechanism that can be used to compute a numerical rating for each team, let's assign some variables to various quantities in question. First, settle on some attribute or statistic of the competition or sport under consideration that you think will be a good basis for making relative comparisons of team strength, and let

a_{ij} = the value of the statistic produced by team i when competing against team j.

For example one of the simplest statistics to consider is the number of times that team i has defeated team j during the current playing season. For ties, assign the participants a value of $1/2$ for each time they have tied. In other words, if W_{ij} is the number of times that team i has recently beaten team j, and if T_{ij} is the number of times that teams i and j have tied, then you might set

$$a_{ij} = W_{ij} + \frac{T_{ij}}{2}.$$

If you think that a more relevant attribute that reflects team i's power relative to that of team j is the number of points S_{ij} that i scores against j, then you might define

$$a_{ij} = S_{ij}.$$

If teams i and j play against each other more than once, then S_{ij} is the cumulative number of points scored by i against j.

Rather than game scores or number of victories, you might choose to use other attributes of the competition. For example, if you are an old-school football fan who believes that the stronger football teams are those that excel in the running game, then you may want to set

$$a_{ij} = \text{the number of rushing yards that team } i \text{ accumulates against team } j.$$

Similarly, a contemporary connoisseur of the NFL could be convinced that it's all about passing, in which case

$$a_{ij} = \text{the number of passing yards that team } i \text{ gained against team } j.$$

Regardless of what attribute you select to define your a_{ij}'s you will have to continually update them throughout the playing season (unless of course you are only evaluating the league after all competition is finished). But as you update your a_{ij}'s, you have to decide if the value of an attribute at the beginning of the season should have as much influence as its value near the end of the season—i.e., do you want to weight the a_{ij}'s as a function of time?

All of this flexibility allows for a lot of fiddling and tweaking, and this is what makes building rating and ranking models so much fun. As you can already begin to see, Keener's method is particularly ripe with opportunities for tinkering, and even more such opportunities will emerge in what is to come.

Laplace's Rule of Succession

Regardless of the attribute that you select, the statistics concerning your attribute can rarely be used in their raw form to create a successful rating technique. For example, consider scores S_{ij}. If teams i and j each have a weak defense but a good offense, then they are prone to rack up big scores against each other. On the other hand, if teams p and q each have a strong defense but a weak offense, then their games are likely to be low scoring. Consequently, the large S_{ij} and S_{ji} can have a disproportionate effect in any subsequent rating system compared to the small S_{pq} and S_{qp}. When comparing teams i and j, it is better to take into account the total number of points scored by setting

$$a_{ij} = \frac{S_{ij}}{S_{ij} + S_{ji}}. \tag{4.1}$$

Similar remarks hold for other attributes—e.g., rushing yards, passing yards, turnovers, etc.

Proportions a_{ij} generated by using ratios such as (4.1) may be better than using raw values, but, as Keener points out, we should really be using something like

$$a_{ij} = \frac{S_{ij} + 1}{S_{ij} + S_{ji} + 2}. \tag{4.2}$$

The motivation for this is Laplace's rule of succession [29], and the intuition is that if

$$p_{ij} = \frac{S_{ij}}{(S_{ij} + S_{ji})}, \quad \text{then} \quad 0 \le p_{ij} \le 1,$$

and p_{ij} can be interpreted as the probability that team i will defeat team j in the future. If team i defeated team j by a cumulative score of X to 0 in past games where $X > 0$, then it might be reasonable to conclude that team i is somewhat better than team j (slightly better if X is small or much better if X is large). However, regardless of the value of $X > 0$, we have that

$$p_{ij} = 1 \quad \text{and} \quad p_{ji} = 0. \tag{4.3}$$

This suggests that it is impossible for team j to ever beat team i in the future, which is clearly unrealistic. On the other hand, if (4.2) is interpreted as the probability p_{ij} that team i will defeat team j in the future, then you can see from (4.2) that $0 < p_{ij} < 1$, and if team i defeated team j by a cumulative score of X to 0 in the past, then

$$p_{ij} = \frac{X + 1}{X + 2} \rightarrow \begin{cases} 1/2 & \text{as } X \rightarrow 0, \\ 1 & \text{as } X \rightarrow \infty, \end{cases}$$

which is more reasonable than (4.3). Moreover, if $S_{ij} \approx S_{ji}$ and both are large, then $p_{ij} \approx 1/2$, but as the difference $S_{ij} - S_{ji} > 0$ increases, p_{ij} gets closer to 1, which makes sense. Consequently, using (4.2) is preferred over (4.1).

To Skew or Not to Skew?

Skewing concerns the issue of how to compensate for the "just because we could" situation in which a stronger team mercilessly runs up the score against a weaker opponent to either enhance their own rating or perhaps to just "rub it in." If (4.2) is used, then Keener suggests applying a nonlinear *skewing function* such as

$$h(x) = \frac{1}{2} + \frac{\text{sgn}\{x - (1/2)\}\sqrt{|2x - 1|}}{2} \tag{4.4}$$

to each of the a_{ij}'s to help mitigate differences at the upper and lower ends. In other words, replace a_{ij} by $h(a_{ij})$. The graph of $h(x)$ is shown in Figure 4.1, and, as you can see, this function has the properties that $h(0) = 0$, $h(1) = 1$, and $h(1/2) = 1/2$. Using the $h(a_{ij})$'s in place of the a_{ij}'s has the intended effect of somewhat moderating differences at the upper and lower ranges. Skewing with $h(x)$ also introduces an artificial separation between two raw values near $1/2$, which might be helpful in distinguishing between teams of nearly equal strength.

Once you have seen Keener's skewing function in (4.4) and have played with it a bit, you should be able to construct many other skewing functions of your own. For example, you may want to customize your skewing function so that it is more or less exaggerated in its skewing ability than is $h(x)$, or perhaps you need your function to affect the upper ranges of the a_{ij}'s differently than how it affects the lower ranges. The value of Keener's

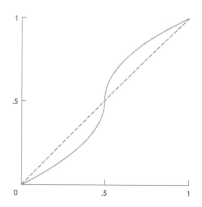

Figure 4.1 Graph of the skewing function $h(x)$

skewing idea is its ability to allow you to "tune" your system to the particular competition being modeled. Skewing is yet another of the countless ways to fiddle with and tweak Keener's technique.

Skewing is not always needed. For example, skewing did not significantly affect the NFL rankings in the studies in [34, 35, 1]. This is probably due to the fact that the NFL was well balanced for the years under consideration, so the need for fine tuning was not so great. Of course, things are generally much different for NCAA sports or other competitions, so don't discount Keener's skewing idea.

Normalization

Having settled on the attribute that defines your nonnegative a_{ij}'s, and having decided whether or not you want to skew them by redefining $a_{ij} \leftarrow h(a_{ij})$, one last bit of massaging (or *normalization*) of the a_{ij}'s is required for situations in which not all teams play the same number of games. In such a case, make the replacement

$$a_{ij} \leftarrow \frac{a_{ij}}{n_i}, \quad \text{where} \quad n_i = \text{the number of games played by team } i. \tag{4.5}$$

To understand why this is necessary, suppose that you are using scores S_{ij} to define your a_{ij}'s along the lines of (4.1) or (4.2). Teams playing more games than other teams have the possibility of producing more points, and thus inducing larger values of a_{ij}. This in turn will affect any measure of "strength" that is eventually derived from your a_{ij}'s. The same is true for statistics other than scores—e.g., if number of yards gained or number of passes completed is your statistic, then teams playing more games can accumulate higher values of these statistics, which in turn will affect any ratings that these statistics produce.

Caution! Skewing has a normalization effect, so if you are using a skewing function and there is not a large discrepancy in the number of games played, then (4.5) may not be necessary—you run the risk of "over normalizing" your data. Furthermore, different situations require different normalization strategies. Using (4.5) is the easiest thing to do and is a good place to start, but you may need to innovate and experiment with other strategies to obtain optimal results—yet another place for fiddling.

Chicken or Egg?

Once the a_{ij}'s are defined, skewed, and normalized, organize them into a square matrix

$$\mathbf{A} = \left[a_{ij}\right]_{m \times m}, \quad \text{where} \quad m = \text{ the number of teams in the league.}$$

Representing the data in such a way allows us to apply some extremely powerful ideas from matrix theory to quantify "strength" and to generate numerical ratings. While it is not apparent at this point, it will be necessary to make a subtle distinction between the "strength value" for each team and its "ratings value." This will become more apparent in the sequel. In spite of the fact that the "strength value" and the "rating value" are not equal, they are of course related. This presents us with a chicken-or-egg situation because we would like to use the "rating values" to measure "strength values" but the "strength" of each team will affect the value of a team's "rating."

Ratings

Let's just jump in blindly and start with ratings. We would like to construct a rating value for each of m teams based on their perceived "strength" (which we still don't have a handle on) at some time t during the current playing season. Let

$$r_j(t) = \text{ the numerical rating of team } j \text{ at time } t,$$

and let $\mathbf{r}(t)$ denote the column containing each of these m ratings. It is understood that the values $r_j(t)$ in the rating vector $\mathbf{r}(t)$ change in time, so the explicit reference to time in the notation can be dropped—i.e., simply write r_j in place of $r_j(t)$ and set

$$\mathbf{r} = \begin{pmatrix} r_1 \\ r_2 \\ \vdots \\ r_m \end{pmatrix} = \textit{the ratings vector.} \tag{4.6}$$

Even though the values in $\mathbf{r}$ are unknown to us at this point, the assumption is that *they exist and can somehow be determined.*

Strength

Now relate the ratings in (4.6) (which at this point exist in theory only) to the concept of "strength." Recall that Keener's first stipulation is that the *strength* of a team should be measured by how well it performs against opponents but tempered by the strength of those opponents. In other words, winning ten games against powder-puff competition shouldn't be considered the same as winning ten games against powerful opponents.

How well team i performs against team j is precisely what your statistic a_{ij} is supposed to measure. How powerful (or powerless) team j is is what the rating value r_j is supposed to gauge. Consequently, it makes sense to adopt the following definition.

Relative Strength

The *relative strength* of team i compared to team j is defined to be

$$s_{ij} = a_{ij}r_j.$$

It is natural to consider the overall or *absolute* strength of team i to be the sum of team i's relative strengths compared to all other teams in the league. In other words, it is reasonable to define *absolute strength* (or simply the *strength*) as follows.

Absolute Strength

The *absolute strength* (or simply the *strength*) of team i is defined to be

$$s_i = \sum_{j=1}^{m} s_{ij} = \sum_{j=1}^{m} a_{ij}r_j, \text{ and } \mathbf{s} = \begin{pmatrix} s_1 \\ s_2 \\ \vdots \\ s_m \end{pmatrix} \text{ is } \textbf{\textit{the strength vector.}} \quad (4.7)$$

Notice that the strength vector $\mathbf{s}$ can be expressed as

$$\mathbf{s} = \begin{pmatrix} \sum_j a_{1j}r_j \\ \sum_j a_{2j}r_j \\ \vdots \\ \sum_j a_{mj}r_j \end{pmatrix} = \begin{pmatrix} a_{11} & a_{12} & \cdots & a_{1m} \\ a_{21} & a_{22} & \cdots & a_{2m} \\ \vdots & \vdots & \ddots & \vdots \\ a_{m1} & a_{m2} & \cdots & a_{mm} \end{pmatrix} \begin{pmatrix} r_1 \\ r_2 \\ \vdots \\ r_m \end{pmatrix} = \mathbf{Ar}. \quad (4.8)$$

The Keystone Equation

Keener's second stipulation concerning the relationship between strength and rating requires that the strength of each team be uniformly proportional to the team's rating in the sense that there is a proportionality constant λ such that $s_i = \lambda r_i$ for each i. In terms of the rating and strength vectors $\mathbf{r}$ and $\mathbf{s}$ in (4.6) and (4.7) this says that $\mathbf{s} = \lambda \mathbf{r}$ for some constant λ. However, (4.8) shows that $\mathbf{s} = \mathbf{Ar}$, so the conclusion is that the ratings vector $\mathbf{r}$ is related to the statistics a_{ij} in matrix $\mathbf{A}$ by means of the equation

$$\mathbf{Ar} = \lambda \mathbf{r}. \quad (4.9)$$

This equation is the keystone of Keener's method!

In the language of linear algebra, equation (4.9) says that the ratings vector $\mathbf{r}$ must be an *eigenvector*, and the proportionality constant λ must be an associated *eigenvalue* for matrix $\mathbf{A}$. Information concerning eigenvalues and eigenvectors can be found in many places on the Internet, but a recommended treatment from a printed source is [54, page 489].

At first glance, it appears that the problem of determining the ratings vector **r** is not only solved, but the solution from (4.9) seems remarkably easy—*simply find the eigenvalues and eigenvectors for some matrix* **A**. However, if determining **r** could not be narrowed down beyond solving a general eigenvalue problem $\mathbf{Ar} = \lambda \mathbf{r}$, then we would be in trouble for a variety of reasons, some of which are listed below.

1. For a general $m \times m$ matrix **A**, there can be as many as m different values of λ that will emerge from the solution (4.9). This would present us with the dilemma of having to pick one value of λ over the others. The choice of λ will in turn affect the ratings vector **r** that is produced because once λ becomes fixed, the ratings vector **r** is married to it as the solution of the equation $(\mathbf{A} - \lambda \mathbf{I})\mathbf{r} = \mathbf{0}$.

2. In spite of the fact that the matrix **A** of statistics contains only real numbers, it is possible that some (or even all) of the λ's that emerge from (4.9) will be complex numbers, which in turn can force the associated eigenvectors **r** to contain complex numbers. Any such **r** is useless for the purpose of rating and ranking anything. For example, it is meaningless for one team to have rating of $6 + 5i$ while another team's rating is $6 - 5i$ because it is impossible to compare these two numbers to decide which is larger.

3. Even in the best of circumstances in which real values of λ emerge from the solution of (4.9), they could be negative numbers. But even if a positive eigenvalue λ pops out, an associated eigenvector **r** can (and usually will) contain some negative entries. While having negative ratings is not as bad as having complex ratings, it is nevertheless not optimal.

4. Finally, you have to face the issue of how you are going to actually compute the eigenvalues and eigenvectors of your matrix so that you can extract **r**. The programming and computational complexity involved in solving $\mathbf{Ar} = \lambda \mathbf{r}$ for a general square matrix **A** of significant size requires most people to buy or have access to a software package that is designed for full-blown eigen computations—most such packages are so expensive that they should be shipped in gold-plated boxes.

Constraints

Keener avoids all of the above stumbling blocks by imposing three mild constraints on the amount of interaction between teams and on the resulting statistics a_{ij} in $\mathbf{A}_{m \times m}$.

I. **Nonnegativity.** Whatever attribute you use to determine the statistics a_{ij}, and however you massage, skew, or normalize them, in the end you must ensure that each statistic a_{ij} is a nonnegative number—i.e., $\mathbf{A} = [a_{ij}] \geq \mathbf{0}$ is a nonnegative matrix.

II. **Irreducibility.** There must be enough past competition within your league to ensure that it is possible to compare any pair of teams, even if they had not played against each other. More precisely, if teams i and j are any two different teams in the league, then they must be "connected" by a a series of past contests involving other teams $\{k_1, k_2, \ldots, k_p\}$ such that there is a string of games

$$i \leftrightarrow k_1 \leftrightarrow k_2 \leftrightarrow \ldots \leftrightarrow k_p \leftrightarrow j \text{ with } a_{ik_1} > 0, \ a_{k_1 k_2} > 0, \ \ldots, \ a_{k_p j} > 0. \quad (4.10)$$

The technical name for this constraint is to say that the competition (or the matrix **A**) is *irreducible*.

III. **Primitivity.** This is just a more stringent version of the irreducibility constraint in II in that we now require each pair of teams to be connected by a *uniform* number of games. In other words, the connection between teams i and j in (4.10) involves a chain of p games with positive statistics. For a different pair of teams, II only requires that they be connected, but they could be connected by a chain of q games with positive statistics in which $p \neq q$. *Primitivity* requires that all teams must be connected by the *same* number of games—i.e., there must exist a single value p such that (4.10) holds for *all* i and j. The primitivity constraint is equivalent to insisting that $\mathbf{A}^p > \mathbf{0}$ for some power p. The reason to require primitivity is because it makes the eventual computation of the ratings vector **r** easy to execute.

Perron–Frobenius

By imposing constraints I and II described above, the powerful Perron–Frobenius theory can be brought to bear to extract a unique ratings vector from Keener's keystone equation $\mathbf{Ar} = \lambda\mathbf{r}$. There is much more to the complete theory than can be presented here (and more than is needed), but its essence is given below—the complete story is given in [54, page 661].

Perron–Frobenius Theorem

If $\mathbf{A}_{m \times m} \geq \mathbf{0}$ is irreducible, then each of the following is true.

- Among all values of λ_i and associated vectors $\mathbf{x}_i \neq \mathbf{0}$ that satisfy $\mathbf{Ax}_i = \lambda_i \mathbf{x}_i$ there is a value λ and a vector $\mathbf{x}$ for which $\mathbf{Ax} = \lambda\mathbf{x}$ such that
  - ▷ λ is real.   ▷ $\lambda > 0$.
  - ▷ $\lambda \geq |\lambda_i|$ for all i.   ▷ $\mathbf{x} > 0$.

- Except for positive multiples of $\mathbf{x}$, there are no other nonnegative eigenvectors $\mathbf{x}_i$ for $\mathbf{A}$, regardless of the eigenvalue λ_i.

- There is a unique vector $\mathbf{r}$ (namely $\mathbf{r} = \mathbf{x}/\sum_j x_j$) for which

$$\mathbf{Ar} = \lambda\mathbf{r}, \quad \mathbf{r} > 0, \quad \text{and} \quad \sum_{j=1}^{m} r_j = 1. \quad (4.11)$$

- The value λ and the vector $\mathbf{r}$ are respectively called the **Perron value** and the **Perron vector**. For us, the Perron value λ is the proportionality constant in (4.9), and the unique Perron vector $\mathbf{r}$ becomes our **ratings vector**.

Important Properties

Notice that by virtue of being defined by the Perron vector for $\mathbf{A}$, the ratings vector $\mathbf{r}$ is not only uniquely determined by the statistics a_{ij} in $\mathbf{A}$, but $\mathbf{r}$ also has the properties that each team rating is such that

$$0 < r_i < 1 \quad \text{and} \quad \sum_{i=1}^{m} r_i = 1.$$

This is important because it puts everything on a level playing field in the sense that it allows for an interpretation of strength in terms of percentages—e.g., a nice pie chart can be made to graphically illustrate the strength of one team relative to the rest of the league. Having the ratings sum to one means that whenever the ratings for a particular team increases, it is necessary for one or more ratings of other teams to decrease, and thus a balance is maintained throughout a given playing season, or from year-to-year, or across different playing seasons. Furthermore, this balance allows level comparisons of ratings derived from two different attributes. For example, if we constructed one set of Keener ratings based on scores S_{ij} and another set of Keener ratings based on wins W_{ij} (and ties), then the two ratings and resulting rankings can be compared, reconciled, and even aggregated to produce what are sometimes called *consensus* (or *ensemble*) ratings and rankings. Techniques of rank aggregation are discussed in detail in Chapter 14.

Computing the Ratings Vector

Given that you have ensured that all statistics a_{ij} are nonnegative (i.e., $\mathbf{A} \geq \mathbf{0}$), then the first thing to do is to check that the irreducibility requirement described in constraint II is satisfied. Let's assume that it is. If it isn't, something will have to be done to fix this, and possible remedies are presented on page 39. Once you are certain that $\mathbf{A}$ is irreducible, then there are two options for computing $\mathbf{r}$.

Brute Force. If you have access to some gold-plated numerical software that has the facility to compute eigenvalues and eigenvectors, then you can just feed your matrix $\mathbf{A}$ into the software and ask it to spit back at you all of the eigenvalues and all of the eigenvectors.

— Sort through the list of eigenvalues that is returned and locate the one that is real, positive, and has magnitude larger than all of the others. The Perron–Frobenius theory ensures that there must be such a value, and this is the Perron root—it is the value of λ that we are looking for.

— Next have your software return an eigenvector $\mathbf{x}$ that is associated with λ. This is *not necessarily* the ratings vector $\mathbf{r}$. Depending on the software that you use, this vector $\mathbf{x} = (x_1, x_2, \ldots, x_m)$ could be returned to you as a vector of all negative numbers or all positive numbers (but no zeros should be present—if they are, something is wrong). Even if $\mathbf{x} > \mathbf{0}$, its components will probably not sum to one, so you generally have to force this to happen by setting

$$\mathbf{r} = \frac{\mathbf{x}}{\sum_{i=1}^{m} x_i}.$$

This is the Perron vector, and consequently $\mathbf{r}$ is the desired ratings vector.

Power Method. If the primitivity condition in III is satisfied, then there is a relatively easy way to compute $\mathbf{r}$ that requires very little in the way of programming skills or computing horse power. It's called the *power method* because it relies on the fact that as k increases, the powers $\mathbf{A}^k$ of $\mathbf{A}$ applied to any positive vector $\mathbf{x}_0$ send the product $\mathbf{A}^k \mathbf{x}_0$ in the direction of the Perron vector—if you are interested in why, see [54, page 533]. Consequently, scaling $\mathbf{A}^k \mathbf{x}_0$ by the sum of its entries and letting $k \to \infty$ produces the Perron vector (i.e., the rating vector) $\mathbf{r}$ as

$$\mathbf{r} = \lim_{k \to \infty} \frac{\mathbf{A}^k \mathbf{x}_0}{\sum_{i=1}^{m} \mathbf{A}^k \mathbf{x}_0}.$$

The computation of $\mathbf{r}$ by the power method proceeds as follows.

— Select an initial positive vector $\mathbf{x}_0$ with which to begin the process. The uniform vector

$$\mathbf{x}_0 = \begin{pmatrix} 1/m \\ 1/m \\ \vdots \\ 1/m \end{pmatrix}$$

is usually a good choice for an initial vector.

— Instead of explicitly computing the powers $\mathbf{A}^k$ and then applying them to $\mathbf{x}_0$, far less arithmetic is required by the following successive calculations. Set

$$\mathbf{y}_k = \mathbf{A}\mathbf{x}_k, \quad \nu_k = \sum_{i=1}^{m} (\mathbf{y}_k)_i, \quad \mathbf{x}_{k+1} = \frac{\mathbf{y}_n}{\nu_k}, \quad \text{for } k = 0, 1, 2, \ldots. \quad (4.12)$$

It is straightforward to verify that this iteration generates the desired sequence

$$\mathbf{x}_1 = \frac{\mathbf{A}\mathbf{x}_0}{\sum_{i=1}^{m} (\mathbf{A}\mathbf{x}_0)_i},$$

$$\mathbf{x}_2 = \frac{\mathbf{A}\mathbf{x}_1}{\sum_{i=1}^{m} (\mathbf{A}\mathbf{x}_1)_i} = \frac{\mathbf{A}^2 \mathbf{x}_0}{\sum_{i=1}^{m} (\mathbf{A}^2 \mathbf{x}_0)_i},$$

$$\mathbf{x}_3 = \frac{\mathbf{A}\mathbf{x}_2}{\sum_{i=1}^{m} (\mathbf{A}\mathbf{x}_2)_i} = \frac{\mathbf{A}^3 \mathbf{x}_0}{\sum_{i=1}^{m} (\mathbf{A}^3 \mathbf{x}_0)_i},$$

$$\vdots$$

The primitivity condition guarantees that

$$\mathbf{x}_k \to \mathbf{r} \quad \text{as} \quad k \to \infty.$$

(Mathematical details are in [54, page 674].) In practice, the iteration (4.12) is terminated when the entries in $\mathbf{x}_k$ have converged to enough significant digits to draw a distinction between the teams in your league. The more teams that you wish to consider, the more significant digits you will need—see the NFL results on page 42 to get a sense of what might be required. Just don't go for overkill. Unless you are trying to rate and rank an inordinately large number of teams, you probably don't need convergence to 16 significant digits.

Forcing Irreducibility and Primitivity

As explained in the discussion of the constraints on page 35, irreducibility and primitivity are conditions that require sufficient interaction between the teams under consideration, and these interactions are necessary for the ratings vector $\mathbf{r}$ to be well defined and to be computable by the power method. This raises the natural question: "How do we actually check to see if our league (or our matrix $\mathbf{A}$) has these kinds of connectivity?"

There is no short cut for checking irreducibility. It boils down to checking the definition—i.e., for each pair of teams (i, j) in your league, you need to verify that there has been a series of games

$$i \leftrightarrow k_1 \leftrightarrow k_2 \leftrightarrow \ldots \leftrightarrow k_p \leftrightarrow j \quad \text{with} \quad a_{ik_1} > 0, \; a_{k_1 k_2} > 0, \; \ldots, \; a_{k_p j} > 0$$

that connects them. Similarly, for primitivity you must verify that there is a *uniform* number of games that connects each pair of teams in your league. These are tedious tasks, but a computer can do the job as long as the number of teams m is not too large.

However, the computational aspect of checking connectivity is generally not the biggest issue because for most competitions it is rare that both kinds of connectivity will be present. This is particularly true when you are building ratings (and rankings) on a continual basis (e.g., week-to-week) from the beginning of a competition and updating them throughout the playing season. It is highly unlikely that you will have sufficient connectivity to ensure irreducibility and primitivity in the early parts of the playing season, and often these connectivity conditions will not be satisfied even near or at the end of a season.

So, the more important question is: "How can we force irreducibility or primitivity into a competition so that we don't ever have to check for it?" One particularly simple solution is to replace $\mathbf{A}$ by a small perturbation of itself by redefining

$$\mathbf{A} \leftarrow \mathbf{A} + \mathbf{E}, \quad \text{where} \quad \mathbf{E} = \epsilon \mathbf{e}\mathbf{e}^T = \begin{pmatrix} \epsilon & \epsilon & \cdots & \epsilon & \epsilon \\ \epsilon & \epsilon & \cdots & \epsilon & \epsilon \\ \vdots & \vdots & \ddots & \vdots & \vdots \\ \epsilon & \epsilon & \cdots & \epsilon & \epsilon \\ \epsilon & \epsilon & \cdots & \epsilon & \epsilon \end{pmatrix}$$

in which $\mathbf{e}$ is a column of all ones and $\epsilon > 0$ is a number that is small relative to the smallest nonzero entry in the unperturbed matrix $\mathbf{A}$. The effect is to introduce an artificial game between every pair of teams such that the statistics of the artificial games are small enough to not contaminate the story that the real games are trying to tell you. Notice that in the perturbed competition every team is now *directly connected* (with positive statistics) to every other team,[1] and therefore both irreducibility and primitivity are guaranteed.

A less stringent perturbation can be used to ensure primitivity when you know that there has been enough games played to guarantee irreducibility. If the competition (or the matrix $\mathbf{A}$) is irreducible, then adding any value $\epsilon > 0$ to any single diagonal position in $\mathbf{A}$ will produce primitivity. In other words, adding one artificial game between some team

---

[1]It is permissible to put zeros in the diagonal positions of $\mathbf{E}$ if having an artificial game between each team and itself bothers you.

and itself with a negligibly small game statistic, or equivalently, redefining $\mathbf{A}$ as

$$\mathbf{A} \leftarrow \mathbf{A} + \mathbf{E}, \quad \text{where} \quad \mathbf{E} = \epsilon e_i e_i^T \quad \text{in which} \quad e_i^T = (0, 0, \ldots, 1, 0, \ldots, 0)$$

will produce primitivity [54, page 678].

Summary

Now that all of the pieces are on the table, let's put them together. Here is a summary of how to build rating and ranking systems from Keener's application of the Perron–Frobenius theory.

1. Begin by choosing one particular attribute of the competition or sport under consideration that you think will be a good basis for making comparisons of each team's strength relative to that of the other teams in the competition. Examples include the number of times team i has defeated (or tied) team j, or the number of points that team i has scored against team j.

   — More than one scheme can be constructed for a given competition by taking more specific features into account—e.g., the number of rushing (or passing) yards that one football team makes against another, or the number of three-point goals or free throws that one basketball team makes against another. Even defensive attributes can be used—e.g., the number of pass interceptions that one football team makes against another, or the number of shots in basketball that team i has blocked when playing team j.

   — The various ratings (and rankings) obtained by using the finer aspects of the competition can be aggregated to form *consensus* (or *ensemble*) ratings (or rankings). For example, you can construct one or more Keener schemes based on offensive attributes and others based on defensive attributes, and then aggregate the resulting ratings into one master rating and ranking list. The details describing how this is accomplished are presented in Chapter 14. Before straying off into rank aggregation, first nail down a solid system that is based on one specific attribute.

2. Whatever attribute is chosen, compile statistics from past competitions, and set

   a_{ij} = the value of the statistic produced by team i when competing against team j.

   It is absolutely necessary that each a_{ij} be a nonnegative number!

3. Massage the raw statistics a_{ij} in step 2 to account for anomalies. For example, if you use $a_{ij} = S_{ij}$ = the number of points that team i scores against team j, then as explained on page 31, you should redefine a_{ij} to be

$$a_{ij} = \frac{S_{ij} + 1}{S_{ij} + S_{ji} + 2} \quad \text{for all } i \text{ and } j.$$

4. If after massaging the data you feel that there is an imbalance in the sense that some a_{ij}'s are very much larger (or smaller) than they should be (perhaps due to a team artificially running up the statistics on a weak opponent), then restore a balance by

constructing a skewing function such as the one in the discussion on page 31 given by

$$h(x) = \frac{1}{2} + \frac{\text{sgn}\{x - (1/2)\}\sqrt{|2x - 1|}}{2},$$

and make the replacement $a_{ij} \leftarrow h(a_{ij})$.

5. If not all teams have played the same number of games, then account for this by normalizing the a_{ij}'s in step 4 above by making the replacement

$$a_{ij} \leftarrow \frac{a_{ij}}{n_i}, \quad \text{where} \quad n_i = \text{the number of games played by team } i.$$

and organize these numbers into a nonnegative matrix $\mathbf{A} = [a_{ij}] \geq \mathbf{0}$.

6. Either check to see that there has been enough competition in your league to guarantee that it (or matrix $\mathbf{A}$) satisfies the irreducibility and primitivity conditions defined on page 36, or else perturb $\mathbf{A}$ by adding some artificial games with negligible game statistics so as to force these conditions as described on page 39.

   — If you are not sure (or just don't want to check) if constraints I and II hold, then both irreducibility and primitivity can be forced by making the replacement $\mathbf{A} \leftarrow \mathbf{A} + \epsilon \mathbf{e}\mathbf{e}^T$, where $\mathbf{e}$ is a column of all ones and $\epsilon > 0$ is small relative to the smallest positive a_{ij} in the unperturbed $\mathbf{A}$.

   — If irreducibility is already present—i.e., if for each pair of teams i and j there has been a series of games such that

   $$i \leftrightarrow k_1 \leftrightarrow k_2 \leftrightarrow \ldots \leftrightarrow k_p \leftrightarrow j \quad \text{with} \quad a_{ik_1} > 0, \; a_{k_1 k_2} > 0, \; \ldots, \; a_{k_p j} > 0$$

   for some p (which can vary with the pair (i, j) being considered)—then the league (or matrix $\mathbf{A}$) can be made primitive (i.e., made to have a uniform p that works for all pairs (i, j)) by simply adding $\epsilon > 0$ to any single diagonal position in $\mathbf{A}$. This creates an artificial game between some team and itself with a negligible statistic, and it amounts to making the replacement $\mathbf{A} \leftarrow \mathbf{A} + \epsilon \mathbf{e}_i \mathbf{e}_i^T$, where $\epsilon > 0$ and $\mathbf{e}_i^T = (0, 0, \ldots, 1, 0, \ldots, 0)$ for some i.

7. Compute the rating vector $\mathbf{r}$ by using the power method described on page 38. If you have forced irreducibility or primitivity as summarized in step 6 above, then slightly modify the power iteration (4.12) on page 38 as shown below.

   — For the perturbed matrix $\mathbf{A} + \epsilon \mathbf{e}\mathbf{e}^T$, execute the power method as follows.

     • Initially set $\mathbf{r} \leftarrow (1/m)\mathbf{e}$, where $\mathbf{e}$ is a column of all ones.
     • Repeat the following steps until the entries in $\mathbf{r}$ converge to a prescribed number of significant digits.

     BEGIN

       1. $\sigma \leftarrow \epsilon \sum_{j=1}^m r_j, \quad (= \epsilon \mathbf{e}^T \mathbf{r})$

       2. $\mathbf{r} \leftarrow \mathbf{A}\mathbf{r} + \sigma \mathbf{e}, \quad (= [\mathbf{A} + \epsilon \mathbf{e}\mathbf{e}^T]\mathbf{r})$

       3. $\nu \leftarrow \sum_{j=1}^m r_j, \quad (= \mathbf{e}^T[\mathbf{A} + \epsilon \mathbf{e}\mathbf{e}^T]\mathbf{r})$

       4. $\mathbf{r} \leftarrow \mathbf{r}/\nu, \quad (= [\mathbf{A} + \epsilon \mathbf{e}\mathbf{e}^T]\mathbf{r}/\mathbf{e}^T[\mathbf{A} + \epsilon \mathbf{e}\mathbf{e}^T]\mathbf{r})$

     REPEAT

— The power method changes a bit when the perturbed matrix $\mathbf{A} + \epsilon \mathbf{e}_i \mathbf{e}_i^T$ is used.

- Initially set $\mathbf{r} \leftarrow (1/m)\mathbf{e}$, where $\mathbf{e}$ is a column of all ones.
- Repeat the following until each entry in $\mathbf{r}$ converges to a prescribed number of significant digits.

BEGIN

1. $\sigma \leftarrow \epsilon r_i,$ $\quad (= \epsilon \mathbf{e}_i^T \mathbf{r})$
2. $\mathbf{r} \leftarrow \mathbf{A}\mathbf{r} + \sigma \mathbf{e}_i,$ $\quad (= [\mathbf{A} + \epsilon \mathbf{e}_i \mathbf{e}_i^T]\mathbf{r})$
3. $\nu \leftarrow \sum_{j=1}^m r_j,$ $\quad (= \mathbf{e}^T[\mathbf{A} + \epsilon \mathbf{e}_i \mathbf{e}_i^T]\mathbf{r})$
4. $\mathbf{r} \leftarrow \mathbf{r}/\nu,$ $\quad (= [\mathbf{A} + \epsilon \mathbf{e}\mathbf{e}^T]\mathbf{r}/\mathbf{e}^T[\mathbf{A} + \epsilon \mathbf{e}_i \mathbf{e}_i^T]\mathbf{r})$

REPEAT

— These steps are simple enough that they can be performed manually in any good spreadsheet environment.

The 2009–2010 NFL Season

To illustrate the ideas in this chapter we used the scores for the regular seventeen-week 2009–2010 NFL season to build Keener ratings and rankings. The teams were ordered alphabetically as shown in the following list of team names.

Order	Name	Order	Name
1.	BEARS	17.	JETS
2.	BENGALS	18.	LIONS
3.	BILLS	19.	NINERS
4.	BRONCOS	20.	PACKERS
5.	BROWNS	21.	PANTHERS
6.	BUCS	22.	PATRIOTS
7.	CARDINALS	23.	RAIDERS
8.	CHARGERS	24.	RAMS
9.	CHIEFS	25.	RAVENS
10.	COLTS	26.	REDSKINS
11.	COWBOYS	27.	SAINTS
12.	DOLPHINS	28.	SEAHAWKS
13.	EAGLES	29.	STEELERS
14.	FALCONS	30.	TEXANS
15.	GIANTS	31.	TITANS
16.	JAGUARS	32.	VIKINGS

Using this ordering, we set

S_{ij} = the cumulative number of points scored by team i against team j

during regular-season games. The matrix containing these raw scores using our ordering is shown below. For example, $S_{12} = 10$ in the following matrix indicates that the BEARS scored 10 points against the BENGALS during the regular season, and $S_{21} = 45$ means that the BENGALS scored 45 points against the BEARS.

	1	2	3	4	5	6	7	8	9	10	11	12	13	14	15	16	17	18	19	20	21	22	23	24	25	26	27	28	29	30	31	32
1	0	10	0	0	30	0	21	0	0	0	0	0	20	14	0	0	0	85	6	29	0	0	0	17	7	0	0	25	17	0	0	46
2	45	0	0	7	39	0	24	17	0	0	0	0	0	0	0	0	0	23	0	31	0	0	17	0	34	0	0	0	41	17	0	10
3	0	0	0	0	3	33	0	0	16	30	0	41	0	3	0	15	29	0	0	0	20	34	0	0	0	0	7	0	0	10	17	0
4	0	12	0	0	27	0	0	37	68	16	17	0	27	0	26	0	0	0	0	3	0	0	20	42	0	0	7	17	0	0	0	0
5	6	27	6	6	0	0	0	23	41	0	0	0	0	0	0	23	0	37	0	3	0	0	23	0	3	0	0	0	27	0	0	20
6	0	0	20	0	0	0	0	0	0	0	21	23	14	27	0	0	3	0	0	38	27	7	0	0	0	13	27	24	0	0	0	0
7	41	0	0	0	0	0	0	0	0	10	0	0	0	0	24	31	0	31	25	7	21	0	0	0	52	0	0	0	58	0	28	30
8	0	27	0	55	30	0	0	0	80	0	20	23	31	0	21	0	0	0	0	0	0	0	48	0	26	23	0	0	28	0	42	0
9	0	10	10	57	34	0	0	21	0	0	20	0	14	0	16	21	0	0	0	0	0	0	0	26	0	24	14	0	0	27	0	0
10	0	0	7	28	0	0	31	0	0	0	0	27	0	0	0	49	15	0	18	0	0	35	0	42	17	0	0	34	0	55	58	0
11	0	0	0	0	0	0	0	0	0	0	0	44	37	55	0	0	0	7	21	0	24	0	0	24	24	38	0	0	0	0	0	0
12	0	0	52	0	0	25	0	13	0	23	0	0	0	7	0	14	61	0	0	0	24	39	0	0	0	0	34	0	24	20	24	0
13	24	0	0	30	0	33	0	23	34	0	16	0	0	34	85	0	0	0	27	0	38	0	9	0	0	54	22	0	0	0	0	0
14	21	0	31	0	0	40	0	0	0	0	21	19	7	0	31	0	10	0	45	0	47	10	0	0	0	31	50	0	0	0	0	0
15	0	0	0	6	0	24	17	20	27	0	64	0	55	34	0	0	0	0	0	0	9	0	44	0	0	68	27	0	0	0	0	7
16	0	0	18	0	17	0	17	0	24	43	0	10	0	0	0	0	24	0	3	0	0	7	0	23	0	0	0	0	0	54	50	0
17	0	37	32	0	0	26	0	0	0	29	0	52	0	7	0	22	0	0	0	0	17	30	38	0	0	0	13	13	0	24	24	0
18	47	13	0	0	38	0	24	0	0	0	0	0	0	0	0	0	0	0	6	12	0	0	0	10	3	19	27	20	20	0	0	23
19	10	0	0	0	0	0	0	44	0	0	14	0	0	13	10	0	20	0	0	24	0	0	0	0	63	0	0	0	40	0	21	27
20	42	24	0	0	31	28	33	0	0	0	17	0	0	0	0	0	0	60	30	0	0	0	0	0	36	27	0	0	48	36	0	49
21	0	0	9	0	0	44	0	0	0	7	17	10	48	41	0	6	0	0	0	0	0	10	0	0	20	43	0	0	0	0	0	26
22	0	0	42	17	0	35	0	0	0	34	0	48	0	26	0	35	40	0	0	0	20	0	0	0	27	0	17	0	0	0	0	0
23	0	20	0	23	9	0	0	36	23	0	7	0	13	0	7	0	0	0	0	0	0	0	0	0	13	13	0	0	27	6	0	0
24	9	0	0	0	0	0	23	0	0	6	0	0	0	0	0	20	0	17	6	17	0	0	0	0	0	7	23	17	0	13	7	31
25	31	21	0	30	50	0	0	31	38	15	0	0	0	0	0	0	48	0	14	0	21	21	0	0	0	0	0	30	0	0	0	0
26	0	0	0	27	0	16	0	20	6	0	6	0	41	17	29	0	0	24	45	0	0	17	0	34	9	0	0	0	0	0	0	0
27	0	0	27	0	0	55	0	0	0	0	17	46	48	61	48	0	24	45	0	0	0	0	40	38	0	28	0	33	0	0	13	9
28	19	0	0	0	0	7	23	0	17	17	0	0	0	0	0	41	0	32	30	10	0	0	0	55	0	0	0	0	0	0	13	27
29	14	32	0	28	33	0	0	38	24	0	30	0	0	0	0	0	28	0	37	0	0	24	0	40	0	0	0	0	0	0	51	0
30	0	28	31	0	0	0	21	0	0	44	0	27	0	0	0	42	7	0	24	0	0	34	29	16	0	47	0	0	0	0	17	10
31	0	0	41	0	0	0	20	17	0	26	0	27	0	0	0	47	17	0	34	0	0	0	0	47	0	0	0	0	34	0	0	0
32	66	30	0	0	34	0	17	0	0	0	0	0	0	0	44	0	0	54	27	68	7	0	0	38	33	0	0	35	17	0	0	0

NFL 2009–2010 regular season scores

From these raw scores we used (4.2) on page 31 to construct the matrix

$$\left[\frac{S_{ij}+1}{S_{ij}+S_{ji}+2}\right]_{32\times32},$$

and then we applied Keener's skewing function $h(x)$ in (4.4) on page 31 to each entry in this matrix to form the nonnegative matrix

$$\mathbf{A}_{32\times32}=[a_{ij}]=\left[h\left(\frac{S_{ij}+1}{S_{ij}+S_{ji}+2}\right)\right].$$

Normalization as described on page 32 is not required because all teams played the same number of games (16 of them—each team had one bye). Furthermore, $\mathbf{A}$ is primitive (and hence irreducible) because $\mathbf{A}^2 > 0$, so there is no need for perturbations. The Perron value for $\mathbf{A}$ is $\lambda \approx 15.832$ and the ratings vector $\mathbf{r}$ (to 5 significant digits) is shown below.

Team	Rating	Team	Rating
BEARS	.029410	JETS	.034683
BENGALS	.031483	LIONS	.025595
BILLS	.029066	NINERS	.031876
BRONCOS	.031789	PACKERS	.035722
BROWNS	.027923	PANTHERS	.030785
BUCS	.026194	PATRIOTS	.035051
CARDINALS	.032346	RAIDERS	.026222
CHARGERS	.035026	RAMS	.024881
CHIEFS	.028006	RAVENS	.033821
COLTS	.034817	REDSKINS	.029107
COWBOYS	.034710	SAINTS	.036139
DOLPHINS	.029805	SEAHAWKS	.027262
EAGLES	.033883	STEELERS	.033529
FALCONS	.032690	TEXANS	.033415
GIANTS	.030480	TITANS	.030538
JAGUARS	.028962	VIKINGS	.034783

(4.13)

NFL 2009–2010 Keener ratings

After the ratings are sorted in descending order, the following ranking of the NFL teams for the 2009–2010 season is produced.

Rank	Team	Rating	Rank	Team	Rating
1.	SAINTS	.036139	17.	BENGALS	.031483
2.	PACKERS	.035722	18.	PANTHERS	.030785
3.	PATRIOTS	.035051	19.	TITANS	.030538
4.	CHARGERS	.035026	20.	GIANTS	.030480
5.	COLTS	.034817	21.	DOLPHINS	.029805
6.	VIKINGS	.034783	22.	BEARS	.029410
7.	COWBOYS	.034710	23.	REDSKINS	.029107
8.	JETS	.034683	24.	BILLS	.029066
9.	EAGLES	.033883	25.	JAGUARS	.028962
10.	RAVENS	.033821	26.	CHIEFS	.028006
11.	STEELERS	.033529	27.	BROWNS	.027923
12.	TEXANS	.033415	28.	SEAHAWKS	.027262
13.	FALCONS	.032690	29.	RAIDERS	.026222
14.	CARDINALS	.032346	30.	BUCS	.026194
15.	NINERS	.031876	31.	LIONS	.025595
16.	BRONCOS	.031789	32.	RAMS	.024881

$$(4.14)$$

NFL 2009–2010 Keener rankings

Most who are familiar with the 2009–2010 NFL season would probably agree that this ranking (and the associated ratings) looks pretty good in that it is an accurate reflection of what happened in the post-season playoffs. The SAINTS ended up being ranked #1, and, in fact, *the* SAINTS *won the Super Bowl*! They defeated the COLTS by a score of 31 to 17. Furthermore, as shown in Figure 5.2 on page 61, the top ten teams in our rankings each ended up making the playoffs during 2009–2010, and this alone adds to the credibility of Keener's scheme.

The COLTS were #5 in our ratings, but they probably would have rated higher if they had not rolled over and deliberately given up their last two games of the season to protect their starters from injury. It might be an interesting and revealing exercise to either delete or somehow weight the scores from the last two regular season games for some (or all) of the teams and compare the resulting ratings and rankings with those above—we have not done this, so if you follow up on your own, then please let us know the outcome.

On the other hand, after reviewing a replay of Super Bowl XLIV, it could be argued that the COLTS actually looked like a #5 team against the SAINTS, and either a healthy PATRIOTS or CHARGERS team might well have given the SAINTS a greater challenge than did the COLTS. Another interesting project would be to weight the scores so that those at the beginning of the season and perhaps some at the end do not count for as much as scores near the crucial period after midseason—and this might be done on a team-by-team basis.

Jim Keener vs. Bill James

Wayne Winston begins his delightful book [83] with a discussion of what is often called the *Pythagorean theorem for baseball* that was formulated by Bill James, a well-known writer, historian, and statistician who has spent much of his time analyzing baseball data. James discovered that the percentage of wins that a baseball team has in one season is closely approximated by a Pythagorean expectation formula that states

$$\% \text{ wins} \approx \frac{\text{runs scored}^2}{\text{runs scored}^2 + \text{runs allowed}^2} = \frac{1}{1 + \rho^2} \quad \text{where} \quad \rho = \frac{\text{runs allowed}}{\text{runs scored}}.$$

This is a widely used idea in the world of quantitative sports analysis—it has been applied to nearly every major sport on the planet. However, each different sport requires a different exponent in the formula. In other words, after picking your sport you must then adapt the formula by replacing ρ^2 with ρ^x, where the value of x is optimized for your sport. Winston suggests using the *mean absolute deviation* (or MAD) for this purpose. That is, if

$$\omega_i = \text{the percentage of games that team } i \text{ wins during a season,} \qquad (4.15)$$

and

$$\rho_i = \frac{\text{points allowed by team } i}{\text{points scored by team } i}, \qquad (4.16)$$

and if there are m teams in the league, then the mean absolute deviation for a given value of x is

$$\text{MAD}(x) = \frac{1}{m} \sum_{i=1}^{m} \left| \omega_i - \frac{1}{1 + \rho_i^x} \right|, \qquad (4.17)$$

or equivalently, in terms of the vector 1-norm [54, page 274],

$$\text{MAD}(x) = \frac{\|\mathbf{w} - \mathbf{p}(x)\|_1}{m}, \quad \text{where} \quad \begin{cases} \mathbf{w} = (\omega_1, \omega_2, \ldots, \omega_m)^T, \\ \text{and} \\ \mathbf{p}(x) = \left((1 + \rho_1^x)^{-1}, (1 + \rho_2^x)^{-1}, \ldots, (1 + \rho_m^x)^{-1} \right)^T. \end{cases}$$

Once you choose a sport then it's your job to find the value x^* that minimizes $\text{MAD}(x)$ for that sport. And if you are really into tweaking things, then "points" in (4.16) can be replaced by other aspects of your sport—e.g., in football it is an interesting exercise to see what happens when you use

$$\rho_i = \frac{\text{yards given up by team } i}{\text{yards gained by team } i}.$$

Let's try out James's Pythagorean idea for the 2009–2010 NFL regular season. In other words, estimate

$$\% \text{ regular season wins} \approx \frac{1}{1 + \rho^x}, \quad \text{where} \quad \rho = \frac{\text{points allowed}}{\text{points scored}},$$

and where x is determined from the 2009–2010 NFL scoring data given on page 43. Then let's compare these results with how well the Keener ratings in (4.13) and (4.14) estimate the winning percentages. The percentage of wins for each NFL team during the regular 2009–2010 season is shown in the following table,[2] and this is our target.

---

[2]For convenience, percent numbers $\star\%$ are converted by multiplying decimal values by 100.

Team	% Wins	Team	% Wins
BEARS	43.75	JETS	56.25
BENGALS	62.50	LIONS	12.50
BILLS	37.50	NINERS	50.00
BRONCOS	50.00	PACKERS	68.75
BROWNS	31.25	PANTHERS	50.00
BUCS	18.75	PATRIOTS	62.50
CARDINALS	62.50	RAIDERS	31.25
CHARGERS	81.25	RAMS	06.25
CHIEFS	25.00	RAVENS	56.25
COLTS	87.50	REDSKINS	25.00
COWBOYS	68.75	SAINTS	81.25
DOLPHINS	43.75	SEAHAWKS	31.25
EAGLES	68.75	STEELERS	56.25
FALCONS	56.25	TEXANS	56.25
GIANTS	50.00	TITANS	50.00
JAGUARS	43.75	VIKINGS	75.00

$$(4.18)$$

Percentage of regular-season wins for the NFL 2009–2010 season

The optimal value x^* of the exponent x in (4.17) is determined by brute force. In other words, compute $\mathrm{MAD}(x)$ for sufficiently many different values of x and then eyeball the results to identify the optimal x^*. Using the 2009–2010 NFL scoring data on page 43 we computed $\mathrm{MAD}(x)$ for 1500 equally spaced values of x between 1 and 4 to estimate

$$x^* = 2.27, \quad \text{and} \quad \mathrm{MAD}(x^*) = .0621. \qquad (4.19)$$

The graph of $\mathrm{MAD}(x)$ is shown in Figure 4.2. This means that the predictions produced by

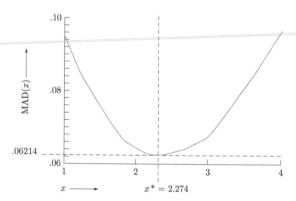

Figure 4.2 Optimal Pythagorean exponent for the NFL

James's generalized Pythagorean formula $1/(1 + \rho^{x^*})$ were only off by an average 6.21% per team for the 2009–2010 NFL season. Not bad for such a simple formula!

Winston reports in [83] that Daryl Morey, the General Manager of the Houston Rockets, made a similar calculation for earlier NFL seasons (it's not clear which ones), and Morey arrived at a Pythagorean exponent of 2.37. Intuition says that the optimal x^* should vary with time. However, Morey's results together with with Winston's calculations and our value suggest that the seasonal variation of x^* for the NFL may be relatively small. Moreover, Winston reports that MAD(2.37) = .061 for the 2005–2007 NFL seasons, which is pretty close to our MAD(2.27) = .062 value for the 2009–2010 season.

Now let's see if the Keener ratings **r** in (4.13) on page 43 can produce equally good results. There is nothing to suggest that Keener's ratings are Pythagorean entities that should work in any kind of nonlinear estimation formula. However, the correlation coefficient between the ratings **r** in (4.13) and the win percentages **w** in (4.18) is

$$R_{\mathbf{rw}} = \frac{(\mathbf{r} - \mu_{\mathbf{r}}\mathbf{e})^T (\mathbf{w} - \mu_{\mathbf{w}}\mathbf{e})}{\|\mathbf{r} - \mu_{\mathbf{r}}\mathbf{e}\|_2 \|\mathbf{w} - \mu_{\mathbf{w}}\mathbf{e}\|_2} \approx .934, \text{ where } \mathbf{e} = \begin{pmatrix} 1 \\ 1 \\ \vdots \\ 1 \end{pmatrix}, \text{ and } \mu_\star = \text{mean}. \quad (4.20)$$

This indicates a strong linear relationship between **r** and the data in **w** [54, page 296] so rather than trying to force the ratings r_i into a Pythagorean paradigm, we should be looking for optimal linear estimates of the form $\alpha + \beta r_i \approx \omega_i$, or equivalently $\alpha \mathbf{e} + \beta \mathbf{r} \approx \mathbf{w}$. The Gauss–Markov theorem [54, page 448] says that "best" values of α and β are those that

minimize $\|\mathbf{Ax} - \mathbf{w}\|_2^2$, where $\mathbf{A} = \begin{pmatrix} 1 & r_1 \\ 1 & r_2 \\ \vdots & \vdots \\ 1 & r_m \end{pmatrix}$, $\mathbf{x} = \begin{pmatrix} \alpha \\ \beta \end{pmatrix}$, and $\mathbf{w} = \begin{pmatrix} \omega_1 \\ \omega_2 \\ \vdots \\ \omega_m \end{pmatrix}$. Finding

α and β boils down to solving the *normal equations* $\mathbf{A}^T\mathbf{Ax} = \mathbf{A}^T\mathbf{w}$ [54, page 226], which is two equations in two unknowns in this case. Solving the normal equations for **x** is a straightforward calculation that is hardwired into most decent hand-held calculators and spreadsheets. For the 2009–2010 NFL data, the solution (to five significant places) is $\alpha = -1.2983$ and $\beta = 57.545$, so the least squares estimate for the percentage of regular season wins based on the Keener (Perron–Frobenius) ratings is (rounded to five places)

$$\text{Est Win\% for team } i = \alpha + \beta r_i = -1.2983 + 57.545\, r_i.$$

The following table shows the actual win percentage (rounded to three places) along with the Pythagorean and Keener estimates for each NFL team in the 2009–2010 regular season.

Team	Wins %	Pythag %	Keener %	Team	Wins %	Pythag %	Keener %
BEARS	43.8	42.2	39.4	JETS	56.3	70.9	69.8
BENGALS	62.5	52.7	51.3	LIONS	12.5	18.9	17.5
BILLS	37.4	36.9	37.5	NINERS	50.0	59.1	53.6
BRONCOS	50.0	50.4	53.1	PACKERS	68.8	73.3	75.7
BROWNS	31.3	27.4	30.9	PANTHERS	50.0	51.3	47.3
BUCS	18.8	24.4	20.9	PATRIOTS	62.5	71.6	71.9
CARDINALS	62.5	58.1	56.3	RAIDERS	31.3	18.2	21.1
CHARGERS	81.3	69.0	71.8	RAMS	6.3	11.0	13.4
CHIEFS	25.0	30.2	31.3	RAVENS	56.3	71.6	64.8
COLTS	87.5	66.7	70.5	REDSKINS	25.0	36.9	37.7
COWBOYS	68.8	69.9	69.9	SAINTS	81.3	71.6	78.1
DOLPHINS	43.8	45.4	41.7	SEAHAWKS	31.3	31.9	27.1
EAGLES	68.8	63.5	65.2	STEELERS	56.3	57.2	63.1
FALCONS	56.3	56.3	58.3	TEXANS	56.3	58.7	62.5
GIANTS	50.0	46.5	45.6	TITANS	50.0	42.8	45.9
JAGUARS	43.8	35.0	36.8	VIKINGS	75.0	71.9	70.3

And now the punch line. The MAD for the Keener estimate is (to three significant places)

$$\text{MAD}_r = \frac{1}{32} \sum_{i=1}^{32} |(-1.2981 + 57.547\, r_i) - \omega_i| = .0591.$$

In other words, by using the Keener ratings to predict the percentage of wins in the 2009–2010 NFL season, we were only off by an average of 5.91% per team, whereas the optimized Pythagorean estimate was off by an average of 6.28% per team. MAD was used because Winston uses it in [83], but the *mean squared error* (or MSE) that averages the sum of squares of the errors is another common metric. For the case at hand we have

$$\text{MSE}_{\text{Pythag}} = .0065 \quad \text{and} \quad \text{MSE}_{\text{Keener}} = .0050.$$

- **Bottom Line.** Keener (Perron–Frobenius really) wins the shootout for 2009–2010!

It is reasonable that Perron–Frobenius (aka James Keener) should be at least as good as Pythagoras (aka Bill James) because the latter does not take into account the level of difficulty (or ease) required for a team to score or to give up points whereas the former does. Is Keener's margin of victory significant? You decide. If your interest is piqued, then conduct more experiments on your own, and let us know what you discover.

Back to the Future

Suppose that you were able to catch a ride with Marty McFly in Dr. Emmett Brown's plutonium powered DeLorean back to the beginning of the 2009–2010 NFL season, and, just like Biff Tannen who took the 2000 edition of *Grays Sports Almanac* back to 1955 to make a fortune, you were able to jot Keener's ratings (4.13) on page 43 on the palm of your hand just before you hopped in next to the flux capacitor. How many winners would you be able to correctly pick by using these ratings? Of course, the real challenge is to try to predict the future and not the past, but the exercise of looking backwards is nevertheless a valuable metric by which to gauge a rating scheme and compare it with others.

Hindsight vs. Foresight

Throughout the book we will refer to the exercise of using current ratings to predict winners of past games as *hindsight prediction*, whereas using current ratings to predict winners of future games is *foresight prediction*. Naturally, we expect hindsight accuracy to be greater than foresight accuracy, but contrary to the popular cliché, hindsight is not 20-20. No single rating value for each team can perfectly reflect all of the past wins and losses throughout the league.

Since most ratings do not account for a home-field advantage, making either hindsight or foresight predictions usually means that you must somehow incorporate a "home-field factor." If you could take the Keener ratings (4.14) on page 44 back to the beginning of the 2009–2010 NFL season to pick winners from all 267 games during the 2009–2010 NFL season, you would have to add a home-field advantage factor of .0008 to the Keener rating for the home team to account for the fact that the average difference between the home-team score and the away-team score was about 2.4.[3] By doing so you would correctly pick 196 out of the 267 games for a winning percentage of about 73.4%. This is the hindsight accuracy.

It is interesting to compare Keener's hindsight predictions with those that might have been obtained had you taken back one of the other ratings that are generally available at the end of the season. There is no shortage of rating sites available on the Internet, but most don't divulge the details involved in determining their ratings. If you are curious how well many of the raters performed for the 2009–2010 NFL season, see Todd Beck's prediction tracker Web site at `www.thepredictiontracker.com/nflresults.php`, where many different raters are compared.

Can Keener Make You Rich?

The Keener ratings are pretty good at estimating the percentage of wins that a given team will have, but if you are a gambler, then that's probably not your primary concern. Making money requires that a bettor has to outfox the bookies, and this usually boils down to beating their point spreads (Chapter 9 on page 113 contains a complete discussion of point spreads and optimal "spread ratings"). The "players" who are reading this have no doubt already asked the question, "How do I use the Keener ratings to predict the spread in a given matchup?" Regrettably, you will most likely lose money if you try to use Keener to estimate point spreads. For starters, the team-by-team scoring differences are poorly correlated with the corresponding differences in the Keener ratings. In fact, when 264 games from the 2009–2010 NFL season were examined, the correlation between differences in game scores and the corresponding differences in the Keener ratings was calculated from (4.20) to be only about .579. In other words, there is not much in the way of a linear relation between the two.

In fact, we would go so far as to say that it is virtually impossible for any single list of rating values to generate accurate estimates of point spreads in American football. General reasons are laid out in detail in Chapter 9 on page 113, but while we are considering Keener ratings, we might ask why they in particular should not give rise to accurate spread estimates. Understanding the answer, requires an appreciation of what is hiding underneath the mathematical sheets. In a nutshell, it is because Keener's technique is the first cousin of a Markov chain defined by making a random walk on the thirty-two NFL teams, and as such, the ratings only convey information about expected wins over a long time horizon. Markovian methods are discussed in detail in Chapter 6 on page 67.

To understand this from an intuitive point of view, suppose that a rabid fan forever travels each week from one NFL city to another and the decision of where to go next week is determined by the entries in the matrix $\mathbf{A} = [a_{ij}]$ that contains your statistics. If you normalize the columns of $\mathbf{A}$ to make each column sum to one, then, just as suggested on page 31, a_{ij} can be thought of as the probability that team i will beat team j. But instead of using this interpretation, you can also think of a_{ij} as being

---

[3]This is not exactly correct because while a "home team" is always designated, three games—namely on 10/25/09 (BUCS–PATRIOTS), 12/3/09 (BILLS–JETS), and 2/7/10 (COLTS–SAINTS)—were played on neutral fields, but these discrepancies are not significant enough to make a substantial difference.

a_{ij} = Probability of traveling to city i given that the fan is currently in city j.

If $a_{ij} > a_{kj}$, then, relative to team j, team i is stronger than k. But in terms of our traveling fan this means that he is destined to spend a greater portion of his life in city i than city k after being in city j. Applying this logic across all cities j suggests that, in the long run, more time is spent in cities with winning teams—the higher the winning percentage, the more time spent. In other words, if we could track our fan's movements to determine exactly what proportion of his existence is spent in each city, then we have in effect gauged the proportion of wins that each team is expected to accumulate in the long run.

The Fan–Keener Connection

The "fan ratings" r_F derived from the proportion of time that a random walking fan spends in each city are qualitatively the same as the Keener ratings r_K. That is, they both reflect the same thing—namely long-term winning percentages.

The intuition is as follows. The sequence $x_0, x_1, x_2, \ldots$ generated by power method (4.12) on page 38 converges to the Keener ratings r_K. Normalizing the Keener matrix at the outset to make each column sum equal one obviates the need to normalize the iterates (i.e., dividing by ν_k in (4.12)) at each step. The power method simplifies to become

$$x_{k+1} = Ax_k \quad \text{for } k = 0, 1, 2, \ldots \quad \text{where} \quad x_0 = \begin{pmatrix} 1/m \\ 1/m \\ \vdots \\ 1/m \end{pmatrix}, \qquad (4.21)$$

and the iterates x_k estimate the path of our fan. The x_0 initiates his trek by giving him an equal probability of starting in any given city. Markov chain theory [54, page 687] ensures that entry i in x_k is the probability that the fan is in city i on week k. If $x_k \to r_F$, then $[r_F]_i$ is the expected percentage of time that the fan spends in city i over an infinite time horizon. The different normalization strategies produce small differences between r_K and r_F,[4] but the results are qualitatively the same. Thus Keener's ratings mirror the fan's expected time in each city, which in turn implies that Keener reflects only long-run win percentages.

Conclusion

Keener's ratings are in essence the result of a limiting process that reflects long-term winning percentages. They say nothing about the short-term scheme of things, and they are nearly useless for predicting point spreads. So, *NO*—Keener's idea won't make you rich by helping you beat the spread, but it was pivotal in propelling two young men into becoming two of the worlds's most wealthy people—if you are curious, then read the following aside.

---

[4]The intuition is valid for NFL data because the variation in the column sums of Keener's A is small, so the alternate normalization strategy amounts to only a small perturbation. The difference between r_K and r_M can be larger when the variation in the column sums in A is larger.

ASIDE: Leaves in the River

An allegory that appeals to us as an applied mathematicians is thinking of mathematical ideas as leaves on the tree of knowledge. A mathematician's job is to shake this tree or somehow dislodge these leaves so that they fall into the river of time. And as they float in and out of the eddies of the ages, many will marvel at their beauty as they gracefully drift by while others will be completely oblivious to them. Some will pluck them out and weave them together with leaves and twigs that have fallen from other branches of science to create larger things that are returned to the river for the benefit and amusement of those downstream.

Between 1907 and 1912 Oskar (or Oscar) Perron (1880–1975) and Ferdinand Georg Frobenius (1849–1917) were shaking the tree to loosen leaves connected to the branch of nonnegative matrices. They had no motives other than to release some of these leaves into the river so that others could appreciate their vibrant beauty. Ranking and rating were nowhere in their consciousness. In fact,

O. Perron   F. G. Frobenius

Perron and Frobenius could not have conceived of the vast array of things into which their leaves would be incorporated. But in relative terms of fame, their theory of nonnegative matrices is known only in rather specialized circles.

The fact that James Keener picked this shiny leaf out of the river in 1993 to develop his rating and ranking system is probably no accident. Keener is a mathematician who specializes in biological applications, and the Perron–Frobenius theory is known to those in the field because it arises naturally in the analysis of evolutionary systems and other aspects of biological dynamics. However, Keener points out in [42] that using Perron–Frobenius ideas for rating and ranking did not originate with him. Leaves

J. Keener   from earlier times were already floating in the river put there by T. H. Wei [81] in 1952, M. G. Kendall [44] in 1955, and again by T. L. Saaty [67] in 1987. Keener's interest in ratings and rankings seems to have been little more than a delightful diversion from his serious science. When his eye caught the colorful glint of this leaf, he lifted it out of the river to admire it for a moment but quickly let it drop back in.

Around 1996–1998 the leaf got caught in a vortex that flowed into **Google.** Larry Page and Sergey Brin, Ph.D. students at Stanford University, formed a collaboration built around their common interest in the World Wide Web. Central to their goal of improving on the existing search engine technology of the time was their development of the **PageRank** algorithm for rating and ranking the importance of Web pages. Again, leaves were lifted from the river and put to use in a fashion similar to

Larry Page   Sergey Brin

that of Keener. While it is not clear exactly which of the Perron–Frobenius colored leaves Brin and Page spotted in the backwaters of the river, it is apparent from PageRank patent documents and early in-house technical reports that their logic paralleled that of Keener and some before him. The mathematical aspects of PageRank and its link to Perron–Frobenius theory can be found in [49]. Brin and Page were able to exploit the rating and ranking capabilities of the Perron vector and weave it together with their Web-crawling software to parlay the result into the megacorporation that is today's **Google.** With a few bright leaves plucked from the river they built a vessel of enormous magnitude that will long remain in the mainstream of the deeper waters.

These stories underscore universal truths about mathematical leaves. When you shake one from the tree into the river, you may be afforded some degree of notoriety if your leaf is noticed a short distance downstream, but you will not become immensely wealthy as a consequence. If your leaf fails to attract attention before drifting too far down the river, its colors become muted, and it ultimately becomes waterlogged and disappears beneath the surface without attracting much attention. However, it is certain that someone else will eventually shake loose similar leaves from the same branch that will attract notice in a different part of the river. And eventually these leaves will be gathered and woven into things of great value.

All of this is to make the point that the river is littered with leaves, many of similar colors and even from the same branch, but fame and fortune frequently elude those responsible for shaking them loose. The tale of Perron–Frobenius, Keener and his predecessors, and the Google guys illustrates how recognition and fortune are afforded less to those who simply introduce an idea and more to those who apply the idea to a useful end.

Historical Note: As alluded to in the preceding paragraphs, using eigenvalues and eigenvectors to develop ratings and rankings has been in practice for a long time—at least sixty years—and this area has a rich history. Some have referred to these methods as "spectral rankings," and Sebastiano Vigna has recently written a nice historical account of the subject. Readers who are interested in additional historical details surrounding spectral rankings are directed to Vigna's paper [79].

By The Numbers —

$\$8,580,000,000 =$ Google's Q1 2011 revenue.

— Has your interest in rating and ranking just increased a bit?

— searchenginewatch.com

Chapter Five

Elo's System

Árpád Élö (1903–1992) was a Hungarian-born physics professor at Marquette University in Milwaukee, Wisconsin. In addition, he was an avid (and excellent) chess player, and this led him to create an effective method to rate and rank chess players. His system was approved by the United States Chess Federation (USCF) in 1960, and by Fédération Internationale des Échecs (the World Chess Federation, or FIDE) in 1970. Elo's idea eventually became popular outside of the chess world, and it has been modified, extended, and adapted to rate other sports and competitive situations. The premise that Elo used was that each chess player's performance is a normally distributed random variable X whose mean μ can change only slowly with time. In other words, while a player

Arpad Elo

might perform somewhat better or worse from one game to the next, μ is essentially constant in the short-run, and it takes a long time for μ to change.

As a consequence, Elo reasoned that once a rating for a player becomes established, then the only thing that can change the rating is the degree to which the player is performing above or below his mean.[1] He proposed a simple linear adjustment that is proportional to the player's deviation from his mean. More specifically, if a player's recent performance (or score [2]) is S, then his old rating $r_{(\text{old})}$ is updated to become his new rating by setting

$$r_{(\text{new})} = r_{(\text{old})} + K(S - \mu), \tag{5.1}$$

where K is a constant—Elo originally set $K = 10$. As more chess statistics became available it was discovered that chess performance is generally not normally distributed, so both the USCF and FIDE replaced the original Elo assumption by one that postulates the expected scoring difference between two players is a logistic function of the difference in their ratings (this is described in detail on page 54). This alteration affects both μ and K in (5.1), but the ratings are still referred to as "Elo ratings."

In 1997 Bob Runyan adapted the Elo system to rate international football (what Americans call soccer), and at some point Jeff Sagarin, who has been providing sports ratings for *USA Today* since 1985, began adapting Elo's system for American football.

---

[1]The assumption that performance is a normally distributed random variable and the idea of defining ratings based on deviations from the mean is found throughout the literature. The paper by Ashburn and Colvert [6] and its bibliography provides statistically inclined readers with information along these lines.

[2]In chess a win is given a score of 1 and a draw is given a score of $1/2$. Performance might be measured by scores from a single match or by scores accumulated during a tournament.

In its present form, the Elo system works like this. You must start with some initial set of ratings (see page 56) for each competitor—to think beyond chess, consider competitors as being teams. Each time teams i and j play against each other, their respective prior ratings $r_{i(\text{old})}$ and $r_{j(\text{old})}$ are updated to become $r_{i(\text{new})}$ and $r_{j(\text{new})}$ by using formulas similar to (5.1). But now everything is going to be considered on a relative basis in the sense that S in (5.1) becomes

$$S_{ij} = \begin{cases} 1 & \text{if } i \text{ beats } j, \\ 0 & \text{if } i \text{ loses to } j, \\ 1/2 & \text{if } i \text{ and } j \text{ tie,} \end{cases} \tag{5.2}$$

and μ becomes

μ_{ij} = the number of points that team i is expected to score against team j.

The new assumption is that μ_{ij} is a logistic function of the difference in ratings

$$d_{ij} = r_{i(\text{old})} - r_{j(\text{old})}$$

prior to teams i and j playing a game against each other. The *standard* logistic function is defined to be $f(x) = 1/(1 + e^{-x})$, but chess ratings employ *the base-ten version*

$$L(x) = \frac{1}{1 + 10^{-x}}.$$

The functions $f(x)$ and $L(x)$ are qualitatively the same because $10^{-x} = e^{-x(\ln 10)}$, and their graphs each have the following characteristic s-shape.

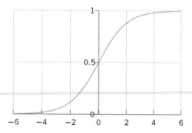

Figure 5.1 Graph of a logistic function

The precise definition of μ_{ij} for Elo chess ratings is

$$\mu_{ij} = L(d_{ij}/400) = \frac{1}{1 + 10^{-d_{ij}/400}}, \quad \text{where} \quad d_{ij} = r_{i(\text{old})} - r_{j(\text{old})}, \tag{5.3}$$

so the respective formulas for updating ratings of teams (or players) i and j are as follows.

Elo Rating Formulas

$$r_{i(\text{new})} = r_{i(\text{old})} + K(S_{ij} - \mu_{ij}) \quad \text{and} \quad r_{j(\text{new})} = r_{j(\text{old})} + K(S_{ji} - \mu_{ji}),$$

where S_{ij} is given by (5.2) and μ_{ij} is given by (5.3).

However, before you can put these formulas to work to build your own rating system, some additional understanding of K and the 400 value in (5.3) is required.

Elegant Wisdom

The simple elegance of the Elo formula belies its wisdom in the sense that Elo inherently rewards a weaker player for defeating a stronger player to greater degree than it rewards a stronger player for beating a weak opponent. For example, if an average chess player's rating is 1500 while that of a stronger player is 1900, then

$$\mu_{avg,str} = \frac{1}{1 + 10^{-(1500-1900)/400}} = \frac{1}{11} \approx .09,$$

but

$$\mu_{str,avg} = \frac{1}{1 + 10^{-(1900-1500)/400}} = \frac{1}{1.1} \approx .91.$$

Therefore, the reward to the average player for beating the stronger player is

$$r_{avg(\text{new})} - r_{avg(\text{old})} = K(S_{avg,str} - \mu_{avg,str}) = K(1 - .09) = .91K,$$

whereas if the stronger player defeats the average player, then the reward is only

$$r_{str(\text{new})} - r_{str(\text{old})} = K(S_{str,avg} - \mu_{str,avg}) = K(1 - .91) = .09K.$$

The K-Factor

The "K-factor," as it is know in chess circles, is still a topic of discussion, and different values are used by different chess groups. Its purpose is to properly balance the deviation between actual and expected scores against prior ratings.

If K is too large, then too much weight is given to the difference between actual and expected scores, and this results in too much volatility in the ratings—e.g., a big K means that playing only a little above expectations can generate a big change in the ratings. On the other hand, if K is too small, then Elo's formulas lose their ability to account for improving or deteriorating play, and the ratings become too stagnant—e.g., a small K means that even a significant improvement in one's play cannot generate much of a change in the ratings.

Chess: The K-factor for chess is often allowed to change with the level of competition. For example, FIDE sets

$K = 25$ for new players until 30 recognized games have been completed;
$K = 15$ for players with > 30 games whose rating has never exceeded 2400;
$K = 10$ for players having reached at least 2400 at some point in the past.

At the time of this writing FIDE is considering changing these respective K-factors to 30, 30, and 20.

Soccer: Some raters of soccer (international football) allow the value of K to increase with the importance of the game. For example, it is not uncommon to see Internet sites report that the following K values are being used.

$K = 60$ for World Cup finals;
$K = 50$ for continental championships and intercontinental tournaments;
$K = 40$ for World Cup qualifiers and major tournaments;
$K = 30$ for all other tournaments;
$K = 20$ for friendly matches.

The K-factor is one of those things that makes building rating systems enjoyable because it allows each rater the freedom to customize his/her system to conform to the specific competition being rated and to add their own personal touch.

The Logistic Parameter ξ

The parameter $\xi = 400$ in the logistic function (5.3) comes from the chess world, and it affects the spread of the ratings. To see what this means in terms of comparing two chess players with respective ratings r_i and r_j, do a bit of algebra to observe that

$$\mu_{ij} = \frac{10^{r_i/400}}{10^{r_i/400} + 10^{r_j/400}} \implies \frac{\mu_{ij}}{\mu_{ji}} = \frac{10^{r_i/400}}{10^{r_j/400}} \implies \mu_{ij} = \mu_{ji} \left[10^{(r_i - r_j)/400} \right].$$

Since μ_{ij} is the number points that player i is expected to score against player j, this means that if $r_i - r_j = 400$, then player i is expected to be ten times better than player j. In general, for every 400 rating points of advantage that player i has over player j, the probability of player i beating player j is expected to be ten times greater than the probability of player j beating player i.

Moving beyond chess, the same analysis as that used above shows that replacing 400 in (5.3) by any value $\xi > 0$ yields

$$\mu_{ij} = L(d_{ij}/\xi) = \frac{1}{1 + 10^{-d_{ij}/\xi}} = \frac{10^{r_i/\xi}}{10^{r_i/\xi} + 10^{r_j/\xi}} \tag{5.4}$$

$$\implies \frac{\mu_{ij}}{\mu_{ji}} = \frac{10^{r_i/\xi}}{10^{r_j/\xi}} \implies \mu_{ij} = \mu_{ji} \left[10^{(r_i - r_j)/\xi} \right].$$

Now, for every ξ rating points of advantage that team i has over team j, the probability of team i beating team j is expected to be ten times greater than the probability of team j beating team i. Fiddling with ξ is yet another way to tune the system to make it optimal for your particular needs.

Constant Sums

When scores S_{ij} depend only on wins, loses, and ties as described in (5.2) for chess, then $S_{ij} + S_{ji} = 1$, and (as shown below) this will force the Elo ratings to always sum to a constant value, regardless of how many times ratings are updated. This constant-sum feature also holds for competitions in which S_{ij} depends on numbers of points scored provided $S_{ij} + S_{ji} = 1$. Below is the formal statement and proof.

The Elo Sums Remain Constant

As long as $S_{ij} + S_{ji} = 1$, then regardless of how the scores S_{ij} are defined, the sum of all Elo ratings $r_i(t)$ at any time $t > 0$ is always the same as the sum of the initial ratings $r_i(0)$. In other words, if there are m teams, then

$$\sum_{k=1}^{m} r_k(0) = \sigma \implies \sum_{k=1}^{m} r_k(t) = \sigma \quad \text{for all } t > 0.$$

In particular, if no competitor deserves an initial bias, then you might assign each participant an initial rating of $r_i(0) = 0$. Doing so guarantees that subsequent Elo ratings always sum to zero, and thus the mean rating is always zero.

Clearly, setting each initial rating equal to x insures that the mean rating at any point in time is always x.

Proof. Notice that regardless of the value of the logistic parameter ξ, equation (5.4) ensures that

$$\mu_{ij} + \mu_{ji} = 1.$$

This together with $S_{ij} + S_{ji} = 1$ means that after a game between teams i and j, the respective updates (from page 54) to the old ratings $r_{i(\text{old})}$ and $r_{j(\text{old})}$ are $K(S_{ij} - \mu_{ij})$ and $K(S_{ji} - \mu_{ji})$, so

$$K(S_{ij} - \mu_{ij}) + K(S_{ji} - \mu_{ji}) = 0. \tag{5.5}$$

Therefore,

$$\sum_{k=1}^{m} r_{k(\text{new})} = \sum_{k \neq i,j}^{m} r_{k(\text{old})} + [r_{i(\text{old})} + K(S_{ij} - \mu_{ij})] + [r_{j(\text{old})} + K(S_{ji} - \mu_{ji})]$$

$$= \sum_{k=1}^{m} r_{k(\text{old})}. \quad \blacksquare$$

Caution! Equation (5.5) says that the *change* in the rating for team i is just the negative of that for team j, but this does *not* mean that $r_{i(\text{new})} + r_{j(\text{new})} = 0$.

Elo in the NFL

To see how Elo performs on something other than chess we implemented the basic Elo scheme that depends only on wins and loses to rate and rank NFL teams for the 2009–2010 season. We considered all of the 267 regular-season and playoff games, and we used the same ordering of the teams as shown in the table on page 42.

While we could have used standings from prior years or from preseason games, we started everything off on an equal footing by setting all initial ratings to zero. Consequently the ratings at any point during the season as well as the final ratings sum to zero, and thus their mean value is zero.

Together with the win-lose scoring rule (5.2) on page 54, we used a K-factor of $K = 32$ (a value used on several Internet gaming sites and in line with those of the chess community), but we increased the logistic factor from 400 to $\xi = 1000$ to provide a better spread in the final ratings. The final rankings are not sensitivie to the value of ξ, and our increase in ξ did not significantly alter the final rankings, which are shown below.

Rank	Team	Rating	Rank	Team	Rating
1.	SAINTS	173.66	17.	PANTHERS	11.474
2.	COLTS	170.33	18.	NINERS	-1.2844
3.	CHARGERS	127.58	19.	GIANTS	-5.3217
4.	VIKINGS	103.50	20.	BRONCOS	-11.126
5.	COWBOYS	89.128	21.	DOLPHINS	-26.717
6.	EAGLES	69.533	22.	BEARS	-28.142
7.	PACKERS	67.829	23.	JAGUARS	-36.214
8.	CARDINALS	53.227	24.	BILLS	-53.350
9.	JETS	50.143	25.	BROWNS	-74.664
10.	PATRIOTS	39.633	26.	RAIDERS	-83.319
11.	TEXANS	33.902	27.	SEAHAWKS	-88.845
12.	BENGALS	33.012	28.	CHIEFS	-109.28
13.	RAVENS	32.083	29.	REDSKINS	-110.21
14.	FALCONS	28.118	30.	BUCS	-130.10
15.	STEELERS	27.125	31.	LIONS	-170.81
16.	TITANS	13.222	32.	RAMS	-194.12

$$(5.6)$$

NFL 2009–2010 Elo rankings based only on wins & loses with $\xi = 1000$ and $K = 32$

Since these Elo ratings are compiled strictly on the basis of wins and loses, it would be surprising if they were not well correlated with the win percentages reported in (4.18) on page 46. Indeed, the correlation coefficient (4.20) is $R_{\mathbf{rw}} = .9921$. Proceeding with linear regression as we did for the Keener ratings on pages 47 and 48 produces

$$\text{Est Win\% for team } i = .5 + .0022268\, r_i,$$

which in turn yields

$$\text{MAD} = .017958 \quad \text{and} \quad \text{MSE} = .0006.$$

While these are good results, it is not clear that they are significant because in essence they boil down to using win-lose information to estimate win percentages.

A more interesting question is, how well does Elo *predict* who wins in the NFL?

Hindsight Accuracy

While it is not surprising that Elo is good at estimating win percentages, it is not clear what to expect if we take the final Elo ratings back in time as described on page 48 to predict in hindsight who wins each game. Looking back at the winners of all of the 267 NFL games during 2009–2010 and comparing these with what Elo predicts—i.e., the higher final rated team is predicted to win—we see that Elo picks 201 game correctly, so

$$\text{Elo Hindsight Accuracy} = 75.3\%.$$

While it seems reasonable to add a home-field-advantage factor to each of the ratings for the home team in order to predict winners, incorporating a home-field-advantage factor does not improve the hindsight accuracy for the case at hand.

As mentioned in the Keener discussion on page 48, the "hindsight is 20-20" cliché is not valid for rating systems—even good ones are not going to predict all games correctly in hindsight. The Elo hindsight accuracy of 75.3% is very good—see the table of comparisons on page 123.

Foresight Accuracy

A more severe test of a ranking system is how well the vector $\mathbf{r}(j)$ of ratings that is computed after game j can predict the winner in game $j + 1$. Of course, we should expect this *foresight* accuracy to be less than the hindsight accuracy because hindsight gets to use complete knowledge of all games played while foresight is only allowed to use information prior to the game currently being analyzed.

A home-field-advantage factor matters much more in Elo foresight accuracy than in hindsight accuracy because if all initial ratings are equal (as they are in our example), then in the beginning, before there is enough competition to change each team's ratings, home-field consideration is the only factor that Elo can use to draw a distinction between two teams. When a home-field advantage factor of $H = 15$ is added to the ratings of the home team when making foresight predictions from the game-by-game ratings computed with $\xi = 1000$ and $K = 32$, Elo correctly predicts the winner in 166 out of 267 games, so

$$\text{Elo Foresight Accuracy} = 62.2\%.$$

Incorporating Game Scores

Instead of building ratings based just on wins and loses, let's jazz Elo up a bit by using the number of points P_{ij} that team i scores against team j in each game. But rather than using raw scores directly, apply the logic that led to (4.2) on page 31 to redefine the "score" that team i makes against team j to be

$$S_{ij} = \frac{P_{ij} + 1}{P_{ij} + P_{ji} + 2}. \tag{5.7}$$

In addition to having $0 < S_{ij} < 1$, this definition ensures that $S_{ij} + S_{ji} = 1$ so that S_{ij} can be interpreted as the probability that team i beats team j (assuming that ties are excluded). Moreover, if all initial ratings are zero (as they are in the previous example), then the discussion on page 56 means that as the Elo ratings change throughout the season, they must always sum to zero and the mean rating is always zero.

When the parameters $\xi = 1000$ and $K = 32$ (the same as those from the basic win-lose Elo scheme) are used with the scores in (5.7), Elo produces the following modified ratings and rankings.

Rank	Team	Rating	Rank	Team	Rating
1.	PACKERS	58.825	17.	BRONCOS	4.1006
2.	VIKINGS	55.217	18.	BENGALS	-.75014
3.	SAINTS	49.495	19.	GIANTS	-3.5097
4.	JETS	47.215	20.	DOLPHINS	-9.3122
5.	COWBOYS	43.074	21.	TITANS	-9.8351
6.	RAVENS	40.357	22.	BEARS	-16.050
7.	CHARGERS	39.974	23.	BILLS	-23.287
8.	COLTS	39.260	24.	REDSKINS	-29.039
9.	PATRIOTS	37.860	25.	CHIEFS	-34.647
10.	NINERS	33.189	26.	SEAHAWKS	-35.150
11.	TEXANS	18.447	27.	JAGUARS	-37.050
12.	FALCONS	18.387	28.	BROWNS	-47.089
13.	EAGLES	13.984	29.	BUCS	-54.373
14.	STEELERS	9.1308	30.	RAIDERS	-62.652
15.	CARDINALS	6.1216	31.	LIONS	-72.800
16.	PANTHERS	5.2596	32.	RAMS	-84.352

$$(5.8)$$

NFL 2009–2010 Elo rankings using scores with $\xi = 1000$ and $K = 32$

Hindsight and Foresight with $\xi = 1000$, $K = 32$, $H = 15$

The ratings in (5.8) produce both hindsight and foresight predictions in a manner similar to the earlier description. But unlike the hindsight predictions produced by considering only wins and losses, the home-field-advantage factor H matters when scores are being considered. If $H = 15$ (empirically determined) is used for both hindsight and foresight predictions, then the ratings in (5.8) correctly predict

194 games (or 72.7%) in hindsight, and 175 games (or 65.5%) in foresight.

Using Variable K-Factors with NFL Scores

The ratings in (5.8) suffer from the defect that they are produced under the assumption that all games are equally important—i.e., the K-factor was the same for all games. However, it can be argued that NFL playoff games should be more important than regular-season games in determining final rankings. Furthermore, there is trend in the NFL for stronger teams to either rest or outright refuse to play their key players to protect them from injuries during the final one or two games of the regular season—this certainly occurred in the final two weeks of the 2009–2010 season.

Consequently, the better teams play below their true scoring ability during this time while their opponents appear to Elo to be stronger than they really are. This suggests that variable K-factors should be used to account for these situations.

Let's try to fix this by setting (rather arbitrarily)

$$K = 32 \quad \text{for the first 15 weeks;}$$
$$K = 16 \quad \text{for the final 2 weeks;}$$
$$K = 64 \quad \text{for all playoff games.}$$

Doing so (but keeping $\xi = 1000$) changes the Elo ratings and rankings to become those shown below in (5.9).

Rank	Team	Rating	Rank	Team	Rating
1.	SAINTS	67.672	17.	CARDINALS	1.4959
2.	VIKINGS	63.080	18.	BENGALS	1.4707
3.	COLTS	57.297	19.	PANTHERS	-3.2548
4.	PACKERS	48.227	20.	DOLPHINS	-7.6586
5.	JETS	38.781	21.	TITANS	-7.7187
6.	CHARGERS	35.864	22.	BEARS	-18.565
7.	RAVENS	35.264	23.	REDSKINS	-22.432
8.	PATRIOTS	28.496	24.	BILLS	-22.709
9.	NINERS	26.047	25.	SEAHAWKS	-29.918
10.	COWBOYS	22.742	26.	JAGUARS	-31.326
11.	TEXANS	16.289	27.	CHIEFS	-35.945
12.	EAGLES	14.492	28.	BROWNS	-51.611
13.	FALCONS	10.531	29.	BUCS	-54.044
14.	STEELERS	7.5351	30.	RAIDERS	-58.546
15.	BRONCOS	7.0388	31.	LIONS	-68.265
16.	GIANTS	6.9994	32.	RAMS	-77.329

(5.9)

NFL 2009–2010 Elo rankings using scores with $\xi = 1000$ and $K = 32, 16, 64$

As far as final rankings are concerned, these in (5.9) look better than those in (5.8) because (5.9) more accurately reflects what actually happened when all the dust had settled. The SAINTS ended up being ranked #1 in (5.9), and, as you can see from the playoff results in Figure 5.2, the SAINTS won the Super Bowl. Furthermore, the VIKINGS gave the SAINTS a greater challenge than the COLTS did during the playoffs, and this showed up in the final ratings and rankings in (5.9).

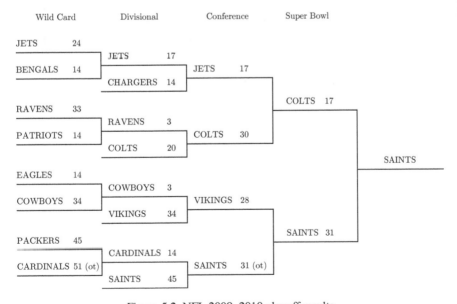

Figure 5.2 NFL 2009–2010 playoff results

Hindsight and Foresight Using Scores and Variable K-Factors

The ratings in (5.9) better reflect the results of the playoffs than those in (5.8), but the ability of (5.9) to predict winners is only slightly better than that of (5.8). Changing the K-factor from a constant $K = 32$ to a variable $K = 32, 16, 64$ changes the value of the ratings, so the home-field-advantage factor H must also be changed—what works best is to use $H = 0$ for hindsight predictions and $H = 9.5$ for foresight predictions. With these values the variable-K Elo system correctly predicts

194 games (or 72.7%) in hindsight,   and   176 games (or 65.9%) in foresight.

Game-by-Game Analysis

Tracking individual team ratings on a game-by-game basis throughout the season provides a sense of overall strength that might not be apparent in the final ratings. For example, when an Elo system that gives more weight to playoff games and discounts the "end-of-season throw-away games" is used (as was done in (5.9)), it is possible that a team may have a higher final ranking than it deserves just because it has a good playoff series. For example, consider the game-by-game ratings for the three highest ranked teams in (5.9). The following three graphs track the game-by-game ratings of the SAINTS vs. COLTS, the SAINTS vs. VIKINGS, and the COLTS vs. VIKINGS.

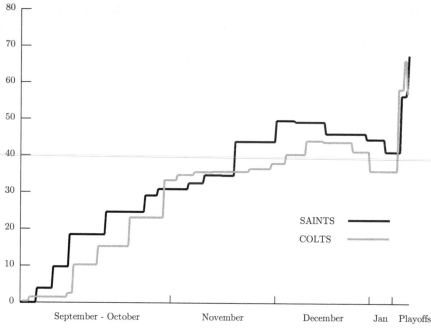

Game-by-game tracking of Elo ratings for SAINTS vs. COLTS

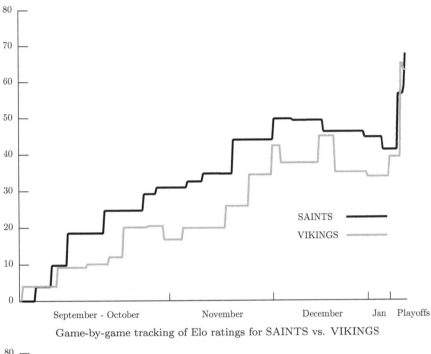

Game-by-game tracking of Elo ratings for SAINTS vs. VIKINGS

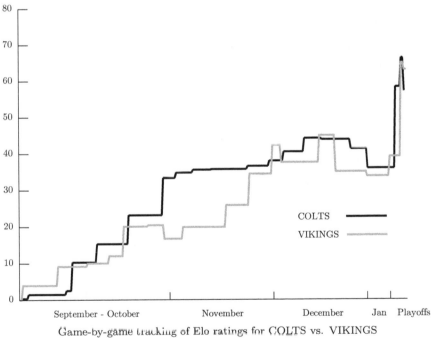

Game-by-game tracking of Elo ratings for COLTS vs. VIKINGS

Notice that the SAINTS are almost always above both the VIKINGS and the COLTS, so this clearly suggests that the SAINTS were the overall stronger team throughout the 2009–2010 season. But how do the COLTS and VIKINGS compare? If the ratings for each are integrated over the entire season (i.e., if the area under the graph of each team in the last graph above is computed), then the COLTS come out ahead of the VIKINGS. In other words, in spite of the fact that VIKINGS are ranked higher than the COLTS in the final ratings, this game-by-game analysis suggests that the COLTS were overall stronger than the VIKINGS across the whole season. As an unbiased fan who saw just about every game that both the COLTS and VIKINGS played during the 2009–2010 season (yes, I [C. M.] have the *Sunday Ticket* and a DVR), I tend to agree with this conclusion.

Conclusion

The Elo rating system is the epitome of simple elegance. In addition, it is adaptable to a wide range of applications. Its underlying ideas and principles might serve you well if you are trying to build a rating system from the ground up.

Regardless of whether or not there was an actual connection between Elo and Facebook, the following aside helps to underscore the adaptability of Elo's rating system.

ASIDE: Elo, Hot Girls, and *The Social Network*

Elo made it into the movies. Well, at least his ranking method did. In a scene from the movie *The Social Network*, Eduardo Saverin (co-founder of Facebook) is entering a dorm room at Harvard University as Mark Zuckerberg (another co-founder of Facebook) is hitting a final keystroke on his laptop, and the dialog goes like this.

Mark: "Done! Perfect timing. Eduardo's here, and he's going to have the key ingredient."

(After comments about Mark's girlfriend troubles)

Eduardo: "Are you alright?"

Mark: "I need you."

Eduardo: "I'm here for you."

Mark: "No, I need the algorithm used to rank chess players."

Eduardo: "Are you okay?"

Mark: "We're ranking girls."

Eduardo: "You mean other students?"

Mark: "Yeah."

Eduardo: "Do you think this is such a good idea?"

Mark: "I need the algorithm—I need the algorithm!"

Eduardo proceeds to write the Elo formula (5.3) from page 54 on their Harvard dorm room window. Well, almost. Eduardo writes the formula as

$$E_a = \frac{1}{1 + 10(R_b - R_a)/400}$$

instead of

$$E_a = \frac{1}{1 + 10^{(R_b - R_a)/400}} = \frac{1}{1 + 10^{-(R_a - R_b)/400}}.$$

We should probably cut the film makers some slack here because, after all, Eduardo is writing on a window with a crayon. The film presents Elo's method as something amazingly complicated that only genius nerds appreciate, which, as you now realize, is not true. Moreover, the film explicitly implies that Elo's method formed the basis for rating Harvard's hot girls on Zuckerberg's Web site Facemash, which was a precursor to Facebook. It makes sense that Zuckerberg might have used Elo's method because Elo is a near-perfect way to rate and rank things by simple "this-or-that" pairwise comparisons. However, probably no one except Zuckerberg himself knows for sure if he actually employed Elo to rate and rank the girls of Harvard. If you know, tell us.

By The Numbers —

2895 = highest chess rating ever.
        — Bobby Fischer (October 1971).

2886 = second highest rating ever.
        — Gary Kasparov (March 1993).

                    — chessbase.com

Chapter Six

The Markov Method

This new method for ranking sports teams invokes an old technique from A. A. Markov, and thus, we call it the Markov method.[1] In 1906 Markov invented his chains, which were later labeled as Markov chains, to describe stochastic processes. While Markov first applied his technique to linguistically analyze the sequence of vowels and consonants in Pushkin's poem *Eugene Onegin*, his chains have found a plethora of applications since [8, 80]. Very recently graduate students of our respective universities, Anjela Govan (Ph.D. North Carolina State University, 2008) [34] and Luke Ingram (M.S., College of Charleston, 2007) [41] used Markov chains to successfully rank NFL football and NCAA basketball teams respectively.

The Markov Method

The Main Idea behind the Markov Method

The Markov rating method can be summarized with one word: *voting*. Every matchup between two teams is an opportunity for the weaker team to vote for the stronger team.

There are many measures by which teams can cast votes. Perhaps the simplest voting method uses wins and losses. A losing team casts a vote for each team that beats it. A more advanced model considers game scores. In this case, the losing team casts a number of votes equal to the margin of victory in its matchup with a stronger opponent. An even more advanced model lets both teams cast votes equal to the number of points given up in the matchup. At the end, the team receiving the greatest number of votes from quality teams earns the highest ranking. This idea is actually an adaptation of the famous PageRank algorithm used by Google to rank webpages [49]. The number of modeling tweaks is nearly endless as the discussion on page 71 shows. However, before jumping to enhancements, let's go directly to our running example.

---

[1]There are at least two other existing sports ranking models that are based on Markov chains. One is the product of Peter Mucha and his colleagues and is known as the Random Walker rating method [17]. The other is the work of Joel Sokol and Paul Kvam and is called LRMC [48].

Voting with Losses

The voting matrix $\mathbf{V}$ that uses only win-loss data appears below.

$$
\mathbf{V} = \begin{array}{c} \\ \text{Duke} \\ \text{Miami} \\ \text{UNC} \\ \text{UVA} \\ \text{VT} \end{array}
\begin{array}{c} \begin{array}{ccccc} \text{Duke} & \text{Miami} & \text{UNC} & \text{UVA} & \text{VT} \end{array} \\
\left(\begin{array}{ccccc}
0 & 1 & 1 & 1 & 1 \\
0 & 0 & 0 & 0 & 0 \\
0 & 1 & 0 & 0 & 1 \\
0 & 1 & 1 & 0 & 1 \\
0 & 1 & 0 & 0 & 0
\end{array} \right) \end{array} .
$$

Because Duke lost to all its opponents, it casts votes of equal weight for each team. In order to tease a ranking vector from this voting matrix we create a (nearly) stochastic matrix $\mathbf{N}$ by normalizing the rows of $\mathbf{V}$.

$$
\mathbf{N} = \begin{array}{c} \\ \text{Duke} \\ \text{Miami} \\ \text{UNC} \\ \text{UVA} \\ \text{VT} \end{array}
\begin{array}{c} \begin{array}{ccccc} \text{Duke} & \text{Miami} & \text{UNC} & \text{UVA} & \text{VT} \end{array} \\
\left(\begin{array}{ccccc}
0 & 1/4 & 1/4 & 1/4 & 1/4 \\
0 & 0 & 0 & 0 & 0 \\
0 & 1/2 & 0 & 0 & 1/2 \\
0 & 1/3 & 1/3 & 0 & 1/3 \\
0 & 1 & 0 & 0 & 0
\end{array} \right) \end{array} .
$$

$\mathbf{N}$ is substochastic at this point because Miami is an undefeated team. This is akin to the well-known dangling node problem in the field of webpage ranking [49]. The solution there, which we borrow here, is to replace all $\mathbf{0}^T$ rows with $1/n\,\mathbf{e}^T$ so that $\mathbf{N}$ becomes a stochastic matrix $\mathbf{S}$.

$$
\mathbf{S} = \begin{array}{c} \\ \text{Duke} \\ \text{Miami} \\ \text{UNC} \\ \text{UVA} \\ \text{VT} \end{array}
\begin{array}{c} \begin{array}{ccccc} \text{Duke} & \text{Miami} & \text{UNC} & \text{UVA} & \text{VT} \end{array} \\
\left(\begin{array}{ccccc}
0 & 1/4 & 1/4 & 1/4 & 1/4 \\
1/5 & 1/5 & 1/5 & 1/5 & 1/5 \\
0 & 1/2 & 0 & 0 & 1/2 \\
0 & 1/3 & 1/3 & 0 & 1/3 \\
0 & 1 & 0 & 0 & 0
\end{array} \right) \end{array} .
$$

While dangling nodes (a webpage with no outlinks) are prevalent in the web context [51, 26], they are much less common over the course of a full sports season. Nevertheless, there are several other options for handling undefeated teams. Consequently, we devote the discussion on page 73 to the presentation of a few alternatives to the uniform row option used above.

Again in analogy with the webpage PageRank algorithm we compute the stationary vector of this stochastic matrix. The stationary vector $\mathbf{r}$ is the dominant eigenvector of $\mathbf{S}$ and solves the eigensystem $\mathbf{Sr} = \mathbf{r}$ [54]. The following short story explains the use of the stationary vector as a means for ranking teams. The main character of the story is a fair weather fan who constantly changes his or her support to follow the current best team. The matrix $\mathbf{S}$ can be represented with the graph in Figure 6.1. The fair weather fan starts at any node in this network and randomly chooses his next team to support based on the links leaving his current team. For example, if this fan starts at UNC and asks UNC, "who is the best team?", UNC will answer "Miami or VT, since they both beat us." The fan flips a coin and suppose he ends up at VT and asks the same question. This time VT answers that Miami is the best team and so he jumps on the Miami bandwagon. Once arriving at the Miami camp he asks his question again. Yet because Miami is undefeated he jumps to any team at random. (This is due to the addition of the $1/5\,\mathbf{e}^T$ row, but there are several other clever strategies for handling undefeated teams that still create a Markov

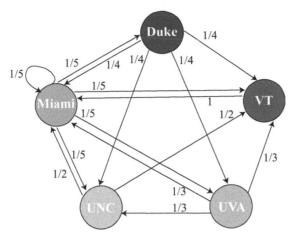

Figure 6.1 Fair weather fan takes a random walk on this Markov graph

chain with nice convergence properties. See page 73.) If the fair weather fan does this continually and we monitor the proportion of time he visits each team, we will have a measure of the importance of each team. The team that the fair weather fan visits most often is the highest ranked team because the fan kept getting referred there by other strong teams. Mathematically, the fair weather fan takes a random walk on a graph defined by a Markov chain, and the long-run proportion of the time he spends in the states of the chain is precisely the stationary vector or dominant eigenvector of the chain. For the five-team example, the stationary rating vector **r** of **S** and its corresponding ranking are in Table 6.1.

Table 6.1 Markov:Losses rating results for the 5-team example

Team	**r**	Rank
Duke	.087	5th
Miami	.438	1st
UNC	.146	3rd
UVA	.110	4th
VT	.219	2nd

Fair Weather Fan Takes a Random Walk on Markov Graph

The Markov rating vector gives the long-run proportion of time that a fair weather fan taking a random walk on the Markov graph spends with each team.

Losers Vote with Point Differentials

Of course, teams may prefer to cast more than one vote for opponents. Continuing with the fair weather fan story, UNC might like to cast more votes for VT than Miami since VT beat them by 27 points while Miami only beat them by 18. In this more advanced model, the voting and stochastic matrices are

$$
\mathbf{V} = \begin{array}{c} \\ \text{Duke} \\ \text{Miami} \\ \text{UNC} \\ \text{UVA} \\ \text{VT} \end{array}
\begin{array}{ccccc} \text{Duke} & \text{Miami} & \text{UNC} & \text{UVA} & \text{VT} \\
\left(\begin{array}{ccccc} 0 & 45 & 3 & 31 & 45 \\ 0 & 0 & 0 & 0 & 0 \\ 0 & 18 & 0 & 0 & 27 \\ 0 & 8 & 2 & 0 & 38 \\ 0 & 20 & 0 & 0 & 0 \end{array}\right) \end{array}
$$

and

$$
\mathbf{S} = \begin{array}{c} \\ \text{Duke} \\ \text{Miami} \\ \text{UNC} \\ \text{UVA} \\ \text{VT} \end{array}
\begin{array}{ccccc} \text{Duke} & \text{Miami} & \text{UNC} & \text{UVA} & \text{VT} \\
\left(\begin{array}{ccccc} 0 & 45/124 & 3/124 & 31/124 & 45/124 \\ 1/5 & 1/5 & 1/5 & 1/5 & 1/5 \\ 0 & 18/45 & 0 & 0 & 27/45 \\ 0 & 8/48 & 2/48 & 0 & 38/48 \\ 0 & 1 & 0 & 0 & 0 \end{array}\right) \end{array}.
$$

This model produces the ratings and rankings listed in Table 6.2. Voting with point differentials results in a slightly different ranking yet significantly different ratings from the elementary voting model. Such differences may become important if one wants to go beyond predicting winners to predict point spreads.

Table 6.2 Markov:Point Differentials rating results for the 5-team example

Team	r	Rank
Duke	.088	5th
Miami	.442	1st
UNC	.095	4th
UVA	.110	3rd
VT	.265	2nd

When two teams faced each other more than once in a season, the modeler has a choice for the entries in the voting matrix. The entry can represent either the cumulative or the average point differential between the two teams in all matchups that season.

Winners and Losers Vote with Points

In a yet more advanced model, we let both the winning and losing teams vote with the number of points given up. This model uses the complete point score information. In this case, the $\mathbf{V}$ and $\mathbf{S}$ matrices (which we now label $\mathbf{V}_{point}$ and $\mathbf{S}_{point}$ to distinguish them from the matrices on page 72) are

$$
\mathbf{V}_{point} = \begin{array}{c} \\ \text{Duke} \\ \text{Miami} \\ \text{UNC} \\ \text{UVA} \\ \text{VT} \end{array}
\begin{array}{ccccc} \text{Duke} & \text{Miami} & \text{UNC} & \text{UVA} & \text{VT} \\
\left(\begin{array}{ccccc} 0 & 52 & 24 & 38 & 45 \\ 7 & 0 & 16 & 17 & 7 \\ 21 & 34 & 0 & 5 & 30 \\ 7 & 25 & 7 & 0 & 52 \\ 0 & 27 & 3 & 14 & 0 \end{array}\right) \end{array}
$$

and

$$\mathbf{S}_{point} = \begin{array}{c} \\ \text{Duke} \\ \text{Miami} \\ \text{UNC} \\ \text{UVA} \\ \text{VT} \end{array} \begin{array}{ccccc} \text{Duke} & \text{Miami} & \text{UNC} & \text{UVA} & \text{VT} \\ \left(\begin{array}{ccccc} 0 & 52/159 & 24/159 & 38/159 & 45/159 \\ 7/47 & 0 & 16/47 & 17/47 & 7/47 \\ 21/90 & 34/90 & 0 & 5/90 & 30/90 \\ 7/91 & 25/91 & 7/91 & 0 & 52/91 \\ 0 & 27/44 & 3/44 & 14/44 & 0 \end{array} \right) \end{array}.$$

Notice that this formulation of $\mathbf{S}$ has the added benefit of making the problem of undefeated teams obsolete. A zero row appears in the matrix only in the unlikely event that a team held all its opponents to scoreless games. The rating and ranking vectors for this model appear in Table 6.3.

Table 6.3 Markov: Winning and Losing Points rating results for the 5-team example

Team	r	Rank
Duke	.095	5th
Miami	.296	1st
UNC	.149	4th
UVA	.216	3rd
VT	.244	2nd

Beyond Game Scores

Sports fans know that game scores are but one of many statistics associated with a contest. For instance, in football, statistics are collected on yards gained, number of turnovers, time of possession, number of rushing yards, number of passing yards, ad nauseam. While points certainly have a direct correlation with game outcomes, this is ultimately not the only thing one may want to predict (think: point spread predictions). Naturally, a model that also incorporates these additional statistics is a more complete model. With the Markov model, it is particularly easy to incorporate additional statistics by creating a matrix for each statistic. For instance, using the yardage information from Table 6.4 below, we build a voting matrix for yardage whereby teams cast votes using the number of yards given up.

Table 6.4 Total yardage data for the 5-team example

	Duke	Miami	UNC	UVA	VT
Duke		100-557	209-357	207-315	35-362
Miami	557-100		321-188	500-452	304-167
UNC	357-209	188-321		270-199	196-338
UVA	315-207	452-500	199-270		334-552
VT	362-35	167-304	338-196	552-334	

$$\mathbf{V}_{yardage} = \begin{array}{c} \\ \text{Duke} \\ \text{Miami} \\ \text{UNC} \\ \text{UVA} \\ \text{VT} \end{array} \begin{pmatrix} \text{Duke} & \text{Miami} & \text{UNC} & \text{UVA} & \text{VT} \\ 0 & 577 & 357 & 315 & 362 \\ 100 & 0 & 188 & 452 & 167 \\ 209 & 321 & 0 & 199 & 338 \\ 207 & 500 & 270 & 0 & 552 \\ 35 & 304 & 196 & 334 & 0 \end{pmatrix}.$$

The (Duke, Miami) element of 577 means that Duke allowed Miami to gain 577 yards against them and thus Duke casts votes for Miami relative to this large number. Row normalizing produces the stochastic matrix $\mathbf{S}_{yardage}$ below.

$$\mathbf{S}_{yardage} = \begin{array}{c} \\ \text{Duke} \\ \text{Miami} \\ \text{UNC} \\ \text{UVA} \\ \text{VT} \end{array} \begin{pmatrix} \text{Duke} & \text{Miami} & \text{UNC} & \text{UVA} & \text{VT} \\ 0 & 577/1611 & 357/1611 & 315/1611 & 362/1611 \\ 100/907 & 0 & 188/907 & 452/907 & 167/907 \\ 209/1067 & 321/1067 & 0 & 199/1067 & 338/1067 \\ 207/1529 & 500/1529 & 270/1529 & 0 & 552/1529 \\ 35/869 & 304/869 & 196/869 & 334/869 & 0 \end{pmatrix}.$$

The rating and ranking vectors produced by $\mathbf{S}_{yardage}$ appear in Table 6.5.

Table 6.5 Markov:Yardage rating results for the 5-team example

Team	r	Rank
Duke	.105	5th
Miami	.249	2nd
UNC	.170	4th
UVA	.260	1st
VT	.216	3rd

Of course, there's nothing special about point or yardage information—any statistical information can be incorporated. The key is to simply remember the voting analogy. Thus, in a similar fashion the $\mathbf{S}_{turnover}$ and $\mathbf{S}_{poss}$ matrices are

$$\mathbf{S}_{turnover} = \begin{array}{c} \\ \text{Duke} \\ \text{Miami} \\ \text{UNC} \\ \text{UVA} \\ \text{VT} \end{array} \begin{pmatrix} \text{Duke} & \text{Miami} & \text{UNC} & \text{UVA} & \text{VT} \\ 0 & 1/9 & 3/9 & 4/9 & 1/9 \\ 3/9 & 0 & 4/9 & 1/9 & 1/9 \\ 2/7 & 3/7 & 0 & 1/7 & 1/7 \\ 1/5 & 0 & 1/5 & 0 & 3/5 \\ 1/10 & 6/10 & 2/10 & 1/10 & 0 \end{pmatrix} \quad \text{and}$$

$$\mathbf{S}_{poss} = \begin{array}{c} \\ \text{Duke} \\ \text{Miami} \\ \text{UNC} \\ \text{UVA} \\ \text{VT} \end{array} \begin{pmatrix} \text{Duke} & \text{Miami} & \text{UNC} & \text{UVA} & \text{VT} \\ 0 & 29.7/118.6 & 30.8/118.6 & 28/118.6 & 30.1/118.6 \\ 30.3/123.6 & 0 & 36.3/123.6 & 31/123.6 & 26/123.6 \\ 29.1/111.6 & 23.7/111.6 & 0 & 27.5/111.6 & 31.3/111.6 \\ 32/131.9 & 29/131.9 & 32.5/131.9 & 0 & 38.4/131.9 \\ 29.9/114.2 & 34/114.2 & 28.7/114.2 & 21.6/114.2 & 0 \end{pmatrix}.$$

The next step is to combine the stochastic matrices for each individual statistic into a single matrix $\mathbf{S}$ that incorporates several statistics simultaneously. As usual, there are a number of ways of doing this. For instance, one can build the lone global $\mathbf{S}$ as a convex combination of the individual stochastic matrices, $\mathbf{S}_{point}, \mathbf{S}_{yardage}, \mathbf{S}_{turnover}$, and $\mathbf{S}_{poss}$ so that

$$\mathbf{S} = \alpha_1 \mathbf{S}_{point} + \alpha_2 \mathbf{S}_{yardage} + \alpha_3 \mathbf{S}_{turnover} + \alpha_4 \mathbf{S}_{poss},$$

where $\alpha_1, \alpha_2, \alpha_3, \alpha_4 \geq 0$ and $\alpha_1 + \alpha_2 + \alpha_3 + \alpha_4 = 1$. The use of the convex combination guarantees that $\mathbf{S}$ is stochastic, which is important for the existence of the rating vector. The weights α_i can be cleverly assigned based on information from sports analysts. However, to demonstrate the model now, we use the simple assignment $\alpha_1 = \alpha_2 = \alpha_3 = \alpha_4 = 1/4$, which means

$$\mathbf{S} = \begin{array}{c} \\ \text{Duke} \\ \text{Miami} \\ \text{UNC} \\ \text{UVA} \\ \text{VT} \end{array} \begin{array}{ccccc} \text{Duke} & \text{Miami} & \text{UNC} & \text{UVA} & \text{VT} \\ \begin{pmatrix} 0 & 0.2617 & 0.2414 & 0.2788 & 0.2182 \\ 0.2094 & 0 & 0.3215 & 0.3055 & 0.1636 \\ 0.2439 & 0.3299 & 0 & 0.1578 & 0.2684 \\ 0.1637 & 0.2054 & 0.1750 & 0 & 0.4559 \\ 0.1005 & 0.4653 & 0.1863 & 0.2479 & 0 \end{pmatrix} \end{array} \quad \text{and} \quad \mathbf{r} = \begin{pmatrix} .15 \\ .24 \\ .19 \\ .20 \\ .21 \end{pmatrix}.$$

The beauty of the Markov model is its flexibility and ability for almost endless tuning. Nearly any piece of statistical information (home-field advantage, weather, injuries, etc.) can be added to the model. Of course, the art lies in the tuning of the α_i parameters.

Handling Undefeated Teams

Some of the methods for filling in entries of the Markov data that were described earlier in this chapter make it very likely to encounter a row of all zeros, which causes the Markov matrix to be substochastic. This substochasticity causes a problem for the Markov model, which requires a row stochastic matrix in order to create a ranking vector. In the example on page 68, we handled the undefeated team of Miami with a standard trick from webpage ranking—replace all $\mathbf{0}^T$ rows with $1/n\,\mathbf{e}^T$. While this mathematically solves the problem and creates a stochastic matrix, there are other ways to handle undefeated teams that may make more sense in the sports setting.

For instance, rather than forcing the best team to cast an equal number of votes for each team, including itself, why not let the best team vote only for itself. In this case, for each undefeated team i, replace the $\mathbf{0}^T$ row with $\mathbf{e}_i^T$, which is a row vector representing the ith row of the identity matrix. Now the data creates a stochastic matrix and hence a Markov chain. However, this fix creates another problem. In fact, this is a potential problem that pertains to the reducibility of the matrix for all Markov models, though it is a certainty in this case. In order for the stationary vector to exist and be unique, the Markov chain must be irreducible (and also aperiodic, which is nearly always satisfied) [54]. An irreducible chain is one in which there is a path from every team to every other team. When undefeated teams vote only for themselves, the chain is reducible as these teams become absorbing states of the chain. In other words, the fair weather fan who takes a random walk on the graph will eventually get stuck at an undefeated team. Here again we employ a trick from webpage ranking and simply add a so-called teleportation matrix $\mathbf{E}$ to the stochasticized voting matrix $\mathbf{S}$ [49]. Thus, for $0 \leq \beta \leq 1$, the Markov matrix $\bar{\mathbf{S}}$ is

$$\bar{\mathbf{S}} = \beta \mathbf{S} + (1 - \beta)/n\,\mathbf{E},$$

where $\mathbf{E}$ is the matrix of all ones and n is the number of teams. Now $\bar{\mathbf{S}}$ is irreducible as every team is directly connected to every other team with at least some small probability, and the stationary vector of $\bar{\mathbf{S}}$ exists and is unique. This rating vector depends on the choice of the scalar β. Generally the higher β is, the truer the model stays to the original data. When the data pertains to the Web, $\beta = .85$ is a commonly cited measure [15]. Yet for sports data, a much lower β is more appropriate. Experiments have shown $\beta = .6$ works well for NFL data [34] and $\beta = .5$ for NCAA basketball [41]. Regardless, the choice for β seems application, data, sport, and even season-specific.

Another technique for handling undefeated teams is to send the fair weather fan back to the team he just came from, where he can then continue his random walk. This bounce-back idea has been implemented successfully in the webpage ranking context [26, 53, 49, 77] where it models nicely the natural reaction of using the BACK button in a browser. While this bounce-back idea is arguably less pertinent in the sports ranking context, it may, nevertheless, be worth exploring. The illustration below summarizes the three methods for handling undefeated teams.

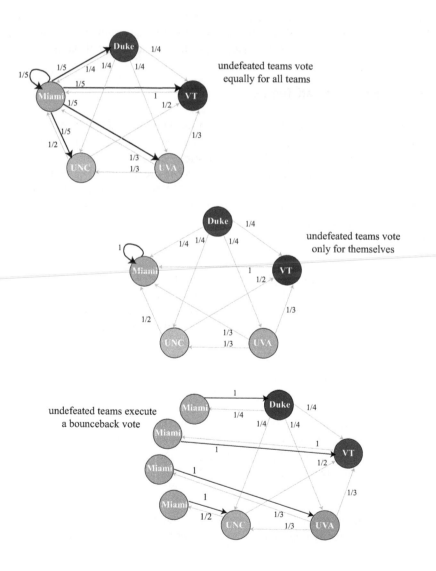

Summary of the Markov Rating Method

The shaded box below assembles the labeling conventions we have adopted to describe our Markov method for ranking sports teams.

Notation for the Markov Rating Method

k number of statistics to incorporate into the Markov model

$\mathbf{V}_{stat1}, \mathbf{V}_{stat2}, \ldots, \mathbf{V}_{statk}$ raw voting matrix for each game statistic k
      $[\mathbf{V}_{stat}]_{ij}$ = number of votes team i casts for team j using statistic $stat$

$\mathbf{S}_{stat1}, \mathbf{S}_{stat2}, \ldots, \mathbf{S}_{statk}$ stochastic matrices built from corresponding voting
      matrices $\mathbf{V}_{stat1}, \mathbf{V}_{stat2}, \ldots, \mathbf{V}_{statk}$

$\mathbf{S}$ final stochastic matrix built from $\mathbf{S}_{stat1}, \mathbf{S}_{stat2}, \ldots, \mathbf{S}_{statk}$;
      $\mathbf{S} = \alpha_1 \mathbf{S}_{stat1} + \alpha_2 \mathbf{S}_{stat2} + \cdots + \alpha_k \mathbf{S}_{statk}$

α_i weight associated with game statistic i; $\sum_{i=1}^{k} \alpha_i = 1$ and $\alpha_i \geq 0$.

$\bar{\mathbf{S}}$ stochastic Markov matrix that is guaranteed to be irreducible;
      $\bar{\mathbf{S}} = \beta \mathbf{S} + (1 - \beta)/n \, \mathbf{E}, \quad 0 < \beta < 1$

$\mathbf{r}$ Markov rating vector; stationary vector (i.e., dominant eigenvector) of $\bar{\mathbf{S}}$

n number of teams in the league = order of $\bar{\mathbf{S}}$

Since most mathematical software programs, such as MATLAB and Mathematica, have built-in functions for computing eigenvectors, the instructions for the Markov method are compact.

MARKOV METHOD FOR RATING TEAMS

1. Form $\mathbf{S}$ using voting matrices for the k game statistics of interest.

$$\mathbf{S} = \alpha_1 \mathbf{S}_{stat1} + \alpha_2 \mathbf{S}_{stat2} + \ldots + \alpha_k \mathbf{S}_{statk},$$

where $\alpha_i \geq 0$ and $\sum_{i=1}^{k} \alpha_i = 1$.

2. Compute $\mathbf{r}$, the stationary vector or dominant eigenvector of $\mathbf{S}$. (If $\mathbf{S}$ is reducible, use the irreducible $\bar{\mathbf{S}} = \beta \mathbf{S} + (1 - \beta)/n \, \mathbf{E}$ instead, where $0 < \beta < 1$.)

We close this section with a list of properties of Markov ratings.

- In a very natural way the Markov model melds many game statistics together.

- The rating vector $\mathbf{r} > \mathbf{0}$ has a nice interpretation as the long-run proportion of time that the fair weather fan (who is taking a random walk on the voting graph) spends with any particular team.

- The Markov method is related to the famous PageRank method for ranking webpages [14].

- Undefeated teams manifest as a row of zeros in the Markov matrix, which makes the matrix substochastic and causes problems for the method. In order to produce a ranking vector, each substochastic row must be forced to be stochastic somehow. The discussion on page 73 outlines several methods for doing this and forces undefeated teams to vote.

- The stationary vector of the Markov method has been shown, like PageRank, to follow a power law distribution. This means that a few teams receive most of the rating, while the great majority of teams must share the remainder of the miniscule leftover rating. One consequence of this is that the Markov method is extremely sensitive to small changes in the input data, particularly when those changes involve low-ranking teams in the tail of the power law distribution. In the sports context, this means that upsets can have a dramatic and sometimes bizarre effect on the rankings. See [21] for a detailed account on this phenomenon.

Connection between the Markov and Massey Methods

While the Markov and Massey methods seem to have little in common, a graphical representation reveals an interesting connection. As pointed out earlier in this chapter, there are several ways to vote in the Markov method. Because the Massey method is built around point differentials, the Markov method that votes with point differentials is most closely connected to the Massey method. Consider Figure 6.2, which depicts the Markov graph

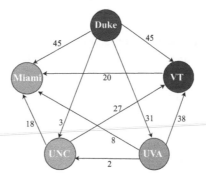

Figure 6.2 Markov graph of point differential voting matrix

associated with the point differential voting matrix V of page 70. This graph actually holds precisely the same information used by the Massey method. Recall the fundamental equation of the Massey method, which for a game k between teams i and j creates a linear equation of the form $r_i - r_j = y_k$, meaning that team i beat team j by y_k points in that matchup. Thus for our five team example, since VT beat UNC by 27 points, Massey creates the equation $r_{VT} - r_{UNC} = 27$. If we create a graph with this point differential information, we have the graph of Figure 6.3, which is simply the transpose of the Markov graph. Both graphs use point differentials to weight the links but the directionality of the links is different. In the Massey graph, the direction of the link indicates dominance whereas it signifies weakness in the Markov graph.

Though the two graphs are essentially the same, the two methods are quite different in philosophy. Given the weights from the links, both methods aim to find weights for the

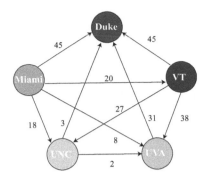

Figure 6.3 Massey graph for the same five team example as Figure 6.2

nodes. The Markov method finds such node weights with a random walk. Recall the fair weather fan who takes a random walk on this graph (or more precisely, the stochasticized version of this graph). The long-run proportion of time the fair weather fan spends at any node i is its node weight r_i. On the other hand, the Massey method finds these node weights with a fitting technique. That is, it finds the node weights r_i that best fit the given link weights. Perhaps the following metaphors are helpful.

Metaphors for the Markov and Massey methods

The Markov method sets a traveler in motion on its graph whereas the Massey method tailors the tightest, most form-fitting outfit to its graph.

ASIDE: PageRank and the Random Surfer

The Markov method of this section was modeled after the Markov chain that Google created to rank webpages. The stationary vector of Google's Markov chain is called the PageRank vector. One way of explaining the meaning of PageRank is with a random surfer, that like our fair weather fan, takes a random walk on a graph. As part of SIAM's Why-doMath? project at www.whydomath.org, a website was created containing a 20-minute video that explains concepts such as PageRank teleportation, dangling node, rank, hyperlink, and random surfer to a non-mathematical audience.

ASIDE: Ranking U.S. Colleges

Each year *US News* publishes a report listing the top colleges in the United States. College marketing departments watch this list carefully as a high ranking can translate into big bucks. Many high school seniors and their parents make their final college decisions based on such rankings. Recently there has been some controversy associated with these rankings. In June 2007, CNN reported that 121 private and liberal arts college administrators were officially pulling their colleges out of the ranking system. The withdrawing institutions called the ranking system unsound because its rankings were built from self-reported data as well as subjective ratings from college presidents and deans.

The controversy surrounding the *US News* college rankings prompted us to explore how these rankings are created. First, the participating colleges and universities are placed into particular categories based on the Carnegie classifications published annually by the Carnegie Foundation for the Advancement of Teaching. For example, the Carnegie classification of Research I applies to all Ph.D.-granting institutions of a certain size. There are similar classifications for Master's I and Master's II schools. Schools are only compared with those in the same Carnegie category. Data are collected about the schools using forms from the Common Data Set developed by *US News*, *Peterson's*, and *The College Board*. Roughly 75% of this information is numerical data, which covers seven major topics: peer assessment, retention and graduation of students, faculty resources, student selectivity, financial resources, alumni giving, and graduation rate. The remaining 25% comes from the subjective ratings of college presidents and deans who are queried about their peer institutions, i.e., schools in their Carnegie classification. *US News* combines this numerical and subjective data using a simple weighted mean whose weights are determined by educational research analysts.

Surprised by the simplicity of the ranking method incorporated by *US News*, we were tempted to compare the *US News* ranking with a more sophisticated ranking method such as the Markov method. For demonstration purposes, we decided to rank fourteen schools that all had the same Carnegie classification. We created a single Markov matrix built from seven matrices based on the following seven factors: number of degrees of study, average institutional financial aid, tuition price, faculty to student ratio, acceptance rate, freshman retention rate, and graduation rate.

Table 6.6 shows both the *US News* ranking (from August 2007) and the Markov ranking for these fourteen schools. Clearly, the two rankings are quite different. Notice, for instance,

Table 6.6 College rankings by *US News* and Markov methods

Institution	*US News* Rank	Markov Rank
Harvard University	1	5
Yale University	2	4
Stanford University	3	12
Columbia University	4	13
University of Chicago	5	11
Brown University	6	14
Emory University	7	10
Vanderbilt University	8	3
University of California, Berkeley	9	7
University of California, Los Angeles	10	9
Brandeis University	11	2
University of Rochester	12	8
University of California, Santa Barbara	13	6
SUNY, Albany	14	1

that SUNY went from last place among its peers in the *US News* ranking to first place in the Markov ranking. Since it is hard to say which ranking is better, our point here is simply that different methods can produce vastly different rankings. And when so much is riding on a ranking (i.e., a school's reputation, and thus, revenue), a sophisticated rigorous method seems preferable.

Chapter Seven

The Offense–Defense Rating Method

A natural approach in the science of ratings is to first rate individual attributes of each team or participant, and then combine these to generate a single number that reflects overall strength. In particular, success in most contests requires a strong offense as well as a strong defense, so it makes sense to try to rate each separately before drawing conclusions about who is #1.

But this is easier said than done, especially with regard to rating offensive strength and defensive strength. The problem is, as everyone knows, playing against a team with an impotent defense can make even the most anemic offense to appear stronger than it really is. And similarly, the apparent defensive prowess of a team is a direct consequence on how strong (or weak) the opposition's offense is. In other words, there is a circular relationship between perceived offensive strength and defensive strength, so each must take the other into account when trying to separately rate offensive power and defensive power. How this is accomplished is the point of this chapter.

OD Objective

The objective is to assign each team (or participant) both an ***offensive rating*** (denoted by $o_i > 0$ for team i) and a ***defensive rating*** (denoted by $d_i > 0$ for team i) that are derived from statistics accumulated from past play. Given any particular sport or contest, there are generally many statistics that might be considered, but scores are the most basic statistics, and actually they do a pretty good job. Moreover, once you understand how our OD method uses scores to produce offensive and defensive ratings, you will then easily see how to use statistics other than scores to build your own systems. For these reasons, we will describe our OD rating technique in terms of scores.

OD Premise

Suppose that m teams are to be rated, and use scoring as the primary measure of performance. Team j is considered to be a good offensive team and deserves a high offensive rating when it scores many points against its opponents, particularly opponents with high defensive ratings. Similarly, team i is a good defensive team and deserves a high defensive rating when it holds its opponents, particularly strong offensive teams, to low scores. Consequently, offensive and defensive ratings are intertwined in the sense that each offensive rating is some function f of all defensive ratings, and each defensive rating is some function g of all offensive ratings.

So, the issue now boils down to determining how the functions f and g should be defined. Since scoring is going to be the primary measure of performance, organize past scores into a **score matrix** $\mathbf{A} = [a_{ij}]$ in which a_{ij} is the score that team j generated against team i (average the results if the teams met more than once, and set $a_{ij} = 0$ if the two teams did not play each other). Notice that a_{ij} has a dual interpretation in the sense that

$$a_{ij} = \begin{cases} \text{Number of points that } j \text{ scores against } i & \text{(an offensive indicator)} \\ \text{and} \\ \text{Number of points that } i \text{ held } j \text{ to} & \text{(a defensive indicator).} \end{cases} \qquad (7.1)$$

In other words, depending on how it is viewed, a_{ij} simultaneously reflects relative offensive and defensive capability. This dual interpretation is used to formulate the following definitions of f and g for each team.

The OD Ratings

For a given set of defensive ratings $\{d_1, d_2, \ldots, d_m\}$, define the **offensive rating** for team j to be

$$o_j = \frac{a_{1j}}{d_1} + \frac{a_{2j}}{d_2} + \cdots + \frac{a_{mj}}{d_m} \quad (\text{large } o_j \implies \text{strong offense}). \qquad (7.2)$$

Similarly, for a given set of offensive ratings $\{o_1, o_2, \ldots, o_m\}$, define the **defensive rating** for team i to be

$$d_i = \frac{a_{i1}}{o_1} + \frac{a_{i2}}{o_2} + \cdots + \frac{a_{im}}{o_m} \quad (\text{large } d_i \implies \text{weak defense}). \qquad (7.3)$$

Reflect on these definitions for a moment to convince yourself that they make sense by reasoning as follows. To understand the offensive rating, suppose that team $\star$ has a weak defense. Consequently, $d_\star$ is relatively large. If team j runs up a big score against team $\star$, then $a_{\star j}$ is large. But this large score is moderated by the large $d_\star$ in the term $a_{\star j}/d_\star$ that appears in the definition of o_j. In other words, scoring a lot of points against a weak defense will not contribute much to the total offensive rating. But if team $\star$ has a strong defense so that $d_\star$ is relatively small, then a large score against $\star$ will not be mitigated by a small $d_\star$, and $a_{\star j}/d_\star$ will significantly increase the offensive rating of team j.

The logic behind the defensive rating is similar. Holding a strong offense to a low score drives the term $a_{i\star}/o_\star$ downward in the defensive rating for team i, but allowing a weak offense to run up the score produces a larger value for $a_{i\star}/o_\star$, which contributes to a larger defensive rating for team i.

But Which Comes First?

Since the OD ratings are circularly defined, it is impossible to directly determine one set without knowing the other. This means that computing the OD ratings can only be accomplished by successive refinement in which one set of ratings is arbitrarily initialized and then successive substitutions are alternately performed to update (7.2) and (7.3) until convergence is realized.

Alternating Refinement Process

Initialize the defensive ratings with positive values stored in a column vector $\mathbf{d}_0$. These initial ratings can be assigned arbitrarily—we will generally set

$$\mathbf{d}_0 = \begin{pmatrix} 1 \\ 1 \\ \vdots \\ 1 \end{pmatrix} = \mathbf{e}.$$

Use the values in $\mathbf{d}_0$ to compute a first approximation to the offensive ratings from (7.2), and store the results in a column vector $\mathbf{o}_1$. Now use these values to compute refinements of the defensive ratings from (7.3), and store the results in a column vector $\mathbf{d}_1$. Repeating this alternating substitution process generates two sequences of vectors

$$\{\mathbf{d}_0, \mathbf{d}_1, \mathbf{d}_2, \mathbf{d}_3, \ldots\} \quad \text{and} \quad \{\mathbf{o}_1, \mathbf{o}_2, \mathbf{o}_3, \mathbf{o}_4, \ldots\}. \tag{7.4}$$

If these sequences converge, then the respective offensive and defensive ratings are the components in the vectors

$$\lim_{k \to \infty} \mathbf{o}_k = \mathbf{o} = \begin{pmatrix} o_1 \\ o_2 \\ \vdots \\ o_m \end{pmatrix} \quad \text{and} \quad \lim_{k \to \infty} \mathbf{d}_k = \mathbf{d} = \begin{pmatrix} d_1 \\ d_2 \\ \vdots \\ d_m \end{pmatrix}.$$

To better analyze this alternating refinement process it is convenient to describe the vectors in this sequence by using matrix algebra along with a bit of nonstandard notation. Given a column vector with positive entries

$$\mathbf{x} = \begin{pmatrix} x_1 \\ x_2 \\ \vdots \\ x_m \end{pmatrix}, \quad \text{define} \quad \mathbf{x}^{\div} = \begin{pmatrix} 1/x_1 \\ 1/x_2 \\ \vdots \\ 1/x_m \end{pmatrix} \quad \text{(the vector of reciprocals)}.$$

If $\mathbf{A} = [a_{ij}]$ is the score matrix defined in (7.1), then the OD ratings in (7.2) and (7.3) are written as the two matrix equations

$$\mathbf{o} = \mathbf{A}^T \mathbf{d}^{\div}, \tag{7.5}$$

and

$$\mathbf{d} = \mathbf{A} \mathbf{o}^{\div}. \tag{7.6}$$

Consequently, with $\mathbf{d}_0 = \mathbf{e}$, the vectors in the OD sequences (7.4) are defined by the alternating refinement process

$$\mathbf{o}_k = \mathbf{A}^T \mathbf{d}_{k-1}^{\div}, \tag{7.7}$$

and

$$\mathbf{d}_k = \mathbf{A} \mathbf{o}_k^{\div}, \quad \text{for } k = 1, 2, 3, \ldots. \tag{7.8}$$

The Divorce

The interdependence between the $\mathbf{o}_k$'s and $\mathbf{d}_k$'s is easily broken so that the computations need not bounce back and forth between offensive and defensive ratings. This is accomplished with a simple substitution of (7.8) into (7.7) to produce the following two independent iterations.

$$\mathbf{o}_k = \mathbf{A}^T \left(\mathbf{A} \mathbf{o}_{k-1}^{\div} \right)^{\div} \quad \text{for } k = 1, 2, 3, \ldots, \tag{7.9}$$

$$\mathbf{d}_k = \mathbf{A} \left(\mathbf{A}^T \mathbf{d}_{k-1}^{\div} \right)^{\div} \quad \text{for } k = 1, 2, 3, \ldots, \tag{7.10}$$

where $\mathbf{o}_0 = \mathbf{e} = \mathbf{d}_0$.

However, there is no need to use both of these iterations because if we can produce ratings from one of them, the other ratings are obtained from (7.5) or (7.6). For example, if the offensive iteration (7.9) converges to $\mathbf{o}$, then plugging it into (7.6) produces $\mathbf{d}$. Similarly, if we only iterate with (7.10), and it converges to $\mathbf{d}$, then $\mathbf{o}$ is produced by (7.5).

Combining the OD Ratings

Once a set of offensive and defensive ratings for each team or participant has been determined, the next problem is to decide how the two sets should be combined to produce a single rating for each team? There are many methods for combining or aggregating rating lists—in fact, Chapters 14 and 15 are devoted to this issue. But for the OD ratings, the simplest (and perhaps the most sensible) thing to do is to define an aggregate or overall rating to be

$$r_i = \frac{o_i}{d_i} \quad \text{for each } i. \tag{7.11}$$

In vector notation, the overall rating vector can be expressed as $\mathbf{r} = \mathbf{o}/\mathbf{d}$, where $/$ denotes component-wise division. Remember that strong offenses are characterized by big o_i's while strong defenses are characterized by small d_i's, so dividing the offensive rating by the corresponding defensive rating makes sense because it allows us to retain the "large value is better" interpretation of the overall ratings in $\mathbf{r}$.

Our Recurring Example

Put the scores in Table 1.1 from our recurring five-team example on page 5 into a scoring matrix

$$
\mathbf{A} = \begin{array}{c} \\ \text{DUKE} \\ \text{MIAMI} \\ \text{UNC} \\ \text{UVA} \\ \text{VT} \end{array}
\begin{array}{c} \text{DUKE} \quad \text{MIAMI} \quad \text{UNC} \quad \text{UVA} \quad \text{VT} \\
\left(\begin{array}{ccccc}
0 & 52 & 24 & 38 & 45 \\
7 & 0 & 16 & 17 & 7 \\
21 & 34 & 0 & 5 & 30 \\
7 & 25 & 7 & 0 & 52 \\
0 & 27 & 3 & 14 & 0
\end{array} \right).
\end{array}
$$

Starting with $\mathbf{d}_0 = \mathbf{e}$ and implementing the alternating iterative scheme in (7.8)–(7.7) requires $k = 9$ iterations to produce convergence of both of the OD vectors to four significant digits. The results are shown in the following table.

O–Rank	Team	O–Rating	D–Rank	Team	D–Rating
1.	MIAMI	151.6	1.	VT	.4104
2.	VT	114.8	2.	MIAMI	.8029
3.	UVA	82.05	3.	UVA	.9675
4.	UNC	48.66	4.	UNC	1.164
5.	DUKE	33.99	5.	DUKE	1.691

OD ratings for the 5-team example

It is clear from these OD ratings and rankings that UVA, UNC, and DUKE should be respectively ranked third, fourth, and fifth overall. However, the OD rankings for VT and MIAMI are flip flopped, so their positions in the overall rankings are not evident. But using the OD ratio in (7.11) resolves the issue. The overall ratings are shown below.

Rank	Team	Rating (o_i/d_i)
1.	VT	279.8
2.	MIAMI	188.8
3.	UVA	84.8
4.	UNC	41.8
5.	DUKE	20.1

(7.12)

Overall ratings for the 5-team example

Scoring vs. Yardage

As mentioned in the introduction to this chapter, almost any statistic derived from a competition could be used in place of scores. For football, it is interesting to see what happens when scores are replaced by yards gained (or given up) in our OD method. As part of his M.S. thesis [41] at the College of Charleston, Luke Ingram used the 2005 ACC Football scoring and yardage data to investigate this issue. The scoring data in our small five-team recurring example is a subset of Luke's experiments. Using a similar subset of Luke's yardage data for the same five teams yields the *yardage matrix*

$$
\mathbf{A} = \begin{array}{c} \\ \text{DUKE} \\ \text{MIAMI} \\ \text{UNC} \\ \text{UVA} \\ \text{VT} \end{array}
\begin{array}{ccccc}
\text{DUKE} & \text{MIAMI} & \text{UNC} & \text{UVA} & \text{VT} \\
\left(\begin{array}{ccccc}
0 & 557 & 357 & 315 & 362 \\
100 & 0 & 188 & 452 & 167 \\
209 & 321 & 0 & 199 & 338 \\
207 & 500 & 270 & 0 & 552 \\
35 & 304 & 196 & 334 & 0
\end{array} \right)
\end{array}.
$$

in which

$$
a_{ij} = \begin{cases} \text{Number of yards team } j \text{ gained against team } i, \\ \text{or equivalently,} \\ \text{Number of yards team } i \text{ gave up against team } j. \end{cases}
$$

Starting with $\mathbf{d}_0 = \mathbf{e}$ and implementing the alternating iterative scheme in (7.8)–(7.7) requires $k = 6$ iterations to produce convergence of both of the OD vectors to four significant digits when the yardage data is used. The results are shown below.

O–Rank	Team	O–Rating	D–Rank	Team	D–Rating
1.	MIAMI	1608	1.	VT	.6629
2.	UVA	1537	2.	MIAMI	.7975
3.	VT	1249	3.	UNC	.9886
4.	UNC	1024	4.	DUKE	1.1900
5.	DUKE	537.3	5.	UVA	1.4020

OD ratings for the 5-team example when yardage is used

This time things are not as clearly defined as when scores are used. The overall ratings obtained from the yardage data is as follows.

Rank	Team	Rating (o_i/d_i)
1.	MIAMI	2016
2.	VT	1885
3.	UVA	1096
4.	UNC	1036
5.	DUKE	451.6

$$(7.13)$$

Overall ratings for the 5-team example when yardage is used

The results in (7.12) are similar to those in (7.13) except that MIAMI and VT have traded places. Experiments tend to suggest that scoring data provides slightly better overall rankings (based on the final standings) for NCAA football than does yardage data. Perhaps this is because the winner is the team with the most points, not necessarily the team with the most total yardage. However, OD ratings based on yardage can be a good indicator of overall strength at particular times in the midst of a season. Of course, results will differ in other sports settings or competitions.

The 2009–2010 NFL OD Ratings

Since we have used the 2009–2010 NFL data in some examples in other places in this book, you might be wondering what the OD ratings look like for the 2009–2010 NFL season. The offensive and defensive ratings were computed using the scoring data for all games (including playoff games), and scores were averaged when two teams played against each other more than once. The results are shown in the following tables.

Rank	Team	O Rating	Rank	Team	O Rating
1.	SAINTS	560.99	17.	TITANS	281.58
2.	VIKINGS	442.37	18.	BRONCOS	256.01
3.	CHARGERS	393.23	19.	CHIEFS	255.6
4.	PACKERS	389.08	20.	PANTHERS	249.47
5.	PATRIOTS	375.69	21.	BENGALS	246.77
6.	COLTS	369.65	22.	NINERS	238.4
7.	EAGLES	362.99	23.	BEARS	231.55
8.	TEXANS	332.85	24.	JAGUARS	224.24
9.	FALCONS	322.99	25.	REDSKINS	219.82
10.	CARDINALS	322.05	26.	BUCS	215.16
11.	RAVENS	315.07	27.	BILLS	214.38
12.	GIANTS	306.6	28.	SEAHAWKS	210.83
13.	DOLPHINS	304.32	29.	LIONS	209.09
14.	STEELERS	304.24	30.	BROWNS	190.22
15.	JETS	298.27	31.	RAIDERS	165.21
16.	COWBOYS	297.71	32.	RAMS	135.81

$$(7.14)$$

Offensive ratings for the 2009–2010 NFL season when scoring is used

Rank	Team	D Rating	Rank	Team	D Rating
1.	JETS	0.6461	17.	PACKERS	1.0391
2.	RAVENS	0.67875	18.	VIKINGS	1.0526
3.	PATRIOTS	0.75685	19.	GIANTS	1.0552
4.	NINERS	0.79343	20.	BEARS	1.056
5.	COWBOYS	0.80562	21.	BUCS	1.0769
6.	PANTHERS	0.81649	22.	RAMS	1.0815
7.	FALCONS	0.85158	23.	TITANS	1.091
8.	BENGALS	0.85385	24.	CHARGERS	1.1005
9.	BILLS	0.87673	25.	SEAHAWKS	1.1363
10.	BRONCOS	0.87783	26.	BROWNS	1.1395
11.	REDSKINS	0.92174	27.	CARDINALS	1.1414
12.	EAGLES	0.94134	28.	RAIDERS	1.1474
13.	COLTS	0.95219	29.	JAGUARS	1.2207
14.	TEXANS	0.95922	30.	CHIEFS	1.2661
15.	STEELERS	1.0113	31.	SAINTS	1.2665
16.	DOLPHINS	1.0143	32.	LIONS	1.4548

$$(7.15)$$

Defensive ratings for the 2009–2010 NFL season when scoring is used

If the ratio $r_i = o_i/d_i$ is used to aggregate the offensive and defensive ratings to produce a single overall rating for each team, then the results for the 2009–2010 NFL season are as follows.

Rank	Team	$r_i = o_i/d_i$	Rank	Team	$r_i = o_i/d_i$
1.	PATRIOTS	496.39	17.	BRONCOS	291.64
2.	RAVENS	464.2	18.	GIANTS	290.56
3.	JETS	461.64	19.	BENGALS	289.01
4.	SAINTS	442.95	20.	CARDINALS	282.15
5.	VIKINGS	420.25	21.	TITANS	258.09
6.	COLTS	388.21	22.	BILLS	244.52
7.	EAGLES	385.61	23.	REDSKINS	238.48
8.	FALCONS	379.28	24.	BEARS	219.26
9.	PACKERS	374.45	25.	CHIEFS	201.89
10.	COWBOYS	369.54	26.	BUCS	199.8
11.	CHARGERS	357.32	27.	SEAHAWKS	185.53
12.	TEXANS	347	28.	JAGUARS	183.7
13.	PANTHERS	305.53	29.	BROWNS	166.94
14.	STEELERS	300.85	30.	RAIDERS	143.98
15.	NINERS	300.46	31.	LIONS	143.73
16.	DOLPHINS	300.03	32.	RAMS	125.57

$$(7.16)$$

Overall ratings for the 2009–2010 NFL season determined by the ratio O/D

However, using this simple ratio is not the only (nor necessarily the best) way to aggregate the OD ratings. For example, it puts as much importance on offense as defense. If you want to experiment with some more complicated strategies, then you might try to build some sort of regression model such as

$$\alpha(o_h - o_w) + \beta(d_h - d_a) = s_h - s_a,$$

where o_h, d_h, o_a, d_a are the respective OD ratings for the home and away teams, and s_h and s_a are the respective scores. **Warning!** Once you start tinkering with ideas like this, you will find that they become infectious. In fact, they are quicksand, and your personal relationships along with your performance at your day job may suffer.

The NFL is not be the best example to illustrate the benefits of our OD method because the NFL is extremely well balanced—much more so than most other sports. The league goes to great lengths to insure this. Consequently, the differentiation between weak and strong offenses (and defenses) is not nearly as pronounced as it is in NCAA competition or in other sports. This means that you won't be too far off the mark by just using raw score totals (without regard to the level of the competition) to gauge offensive and defensive prowess in the NFL, but in other sports or competitions this may not be the case. For example, when the same scoring matrix **A** that produces the NFL OD ratings shown above is used to compute

$$\text{Avg Pts Scored Per Game} = \frac{\text{Total points scored}}{\text{Number of games played}},$$

and

$$\text{Avg Pts Allowed Per Game} = \frac{\text{Total points allowed}}{\text{Number of games played}},$$

the following results are produced.

Rank	Team	Avg Pts Scored	Rank	Team	Avg Pts Scored
1.	SAINTS	33.687	17.	JETS	21.107
2.	PACKERS	30.115	18.	NINERS	19.731
3.	VIKINGS	29.2	19.	BENGALS	19.615
4.	PATRIOTS	27.346	20.	BRONCOS	19.423
5.	EAGLES	27.192	21.	PANTHERS	19.038
6.	CHARGERS	26.893	22.	BEARS	19
7.	RAVENS	25.808	23.	CHIEFS	18.615
8.	COLTS	25.179	24.	JAGUARS	18.042
9.	TEXANS	24.577	25.	REDSKINS	17.538
10.	CARDINALS	24.536	26.	SEAHAWKS	17.385
11.	STEELERS	24.269	27.	LIONS	17
12.	GIANTS	23.731	28.	BUCS	16.958
13.	FALCONS	22.654	29.	BROWNS	16.654
14.	TITANS	22.375	30.	BILLS	15.846
15.	COWBOYS	21.893	31.	RAIDERS	13
16.	DOLPHINS	21.846	32.	RAMS	11.692

$$(7.17)$$

Average points scored per game during the 2009–2010 NFL season

Rank	Team	Avg Pts Allowed	Rank	Team	Avg Pts Allowed
1.	JETS	15.346	17.	STEELERS	20.769
2.	RAVENS	16.077	18.	EAGLES	20.846
3.	COWBOYS	17.357	19.	SAINTS	21.25
4.	BENGALS	18.808	20.	BEARS	22.885
5.	VIKINGS	19.133	21.	CARDINALS	23.643
6.	COLTS	19.25	22.	JAGUARS	23.923
7.	NINERS	19.269	23.	BROWNS	24.154
8.	PANTHERS	19.308	24.	DOLPHINS	24.577
9.	TEXANS	19.462	25.	RAIDERS	24.692
10.	BRONCOS	19.731	26.	TITANS	24.808
11.	FALCONS	19.769	27.	RAMS	25.192
12.	PATRIOTS	19.958	28.	BUCS	25.423
13.	BILLS	20.231	29.	CHIEFS	26.038
14.	REDSKINS	20.231	30.	SEAHAWKS	27.708
15.	PACKERS	20.346	31.	GIANTS	28.542
16.	CHARGERS	20.714	32.	LIONS	30.346

$$(7.18)$$

Average points allowed per game during the 2009–2010 NFL season

If you take a moment to compare the offensive ratings in (7.14) with average points scored in (7.17), you see that while they do not provide exactly the same rankings, they are nevertheless more or less in line with each other—as they should be. Something would be amiss if this were not the case because the same data produced both sets of rankings. The difference is that offensive ratings in (7.14) take the level of competition into account whereas the raw scores in (7.17) do not. Similarly, the defensive ratings (7.15) are more or less in line with the average points allowed shown in (7.18)—all is right in the world.

Mathematical Analysis of the OD Method

The algorithm for computing the OD ratings is really simple—you just alternate the two multiplications (7.7) and (7.8) as described on page 81. Among all of the rating schemes discussed in this book, our OD method is one of the easiest to implement—an ordinary spread sheet can comfortably do this for you, or, if there are not too many competitors, a hand calculator or even pencil-and-paper will do the job. However, the simplicity of implementation belies some of the deeper mathematical issues surrounding the theory of the OD algorithm.

The main theoretical problem concerns the existence of the OD ratings. Since the vectors **o** and **d** that contain the OD ratings are defined to be the limits

$$\mathbf{o} = \lim_{k \to \infty} \mathbf{o}_k \quad \text{and} \quad \mathbf{d} = \lim_{k \to \infty} \mathbf{d}_k,$$

where

$$\mathbf{o}_k = \mathbf{A}^T \mathbf{d}_{k-1}^{\div} \quad \text{and} \quad \mathbf{d}_k = \mathbf{A} \mathbf{o}_k^{\div}, \quad \text{with } \mathbf{d}_0 = \mathbf{e},$$

there is a real possibility that **o** and **d** may not exist—indeed, there are many cases in which these sequences fail to converge.

A theorem is needed to tell us when convergence is guaranteed and when convergence is impossible. Before developing such a theorem, some terminology, notation, and background information are required. Things hinge on the realization that the OD method is closely related to a successive scaling or balancing procedure on the data matrix $\mathbf{A}_{m \times m}$. To understand this, suppose that the columns of $\mathbf{A}$ are first scaled (or balanced) to make all column sums equal to one, and then the resulting matrix is row-scaled to make all row sums equal to one. This can be described as multiplication with *diagonal scaling matrices* $\mathbf{C}_1$ and $\mathbf{R}_1$ such that $\mathbf{AC}_1$ has all column sums equal to one followed by $\mathbf{R}_1(\mathbf{AC}_1)$ to force all row sums equal to one. In other words, if

$$\mathbf{C}_1 = \begin{pmatrix} \xi_1 & 0 & \cdots & 0 \\ 0 & \xi_2 & \cdots & 0 \\ \vdots & \vdots & \ddots & \vdots \\ 0 & 0 & \cdots & \xi_n \end{pmatrix} \quad \text{where} \quad \xi_i = \frac{1}{[e^T \mathbf{A}]_i} \quad \text{(reciprocal of the } i^{th} \text{ column sum)}$$

and

$$\mathbf{R}_1 = \begin{pmatrix} \rho_1 & 0 & \cdots & 0 \\ 0 & \rho_2 & \cdots & 0 \\ \vdots & \vdots & \ddots & \vdots \\ 0 & 0 & \cdots & \rho_n \end{pmatrix} \quad \text{where} \quad \rho_i = \frac{1}{[(\mathbf{AC}_1)e]_i} \quad \text{(reciprocal of the } i^{th} \text{ row sum),}$$

then
$$\mathbf{R}_1 \mathbf{AC}_1 e = e \quad \text{(i.e., all row sums are one).}$$

Unfortunately, scaling the rows of $\mathbf{AC}_1$ destroys the column balance achieved in the previous step when we multiplied $\mathbf{A}$ by $\mathbf{C}_1$—i.e., the column sums of $\mathbf{R}_1 \mathbf{AC}_1$ are generally not equal to one. However, the column sums in $\mathbf{R}_1 \mathbf{AC}_1$ are often closer to being equal to one than those of $\mathbf{A}$. Consequently, the same kind of alternate scaling is performed again, and again, and again, until the data is perfectly balanced in the sense that the row sums as well as the column sums are all equal to one (or at least close enough to one within some prescribed tolerance). Successive column scaling followed by row scaling generates a sequence of diagonal scaling matrices $\mathbf{C}_k$ and $\mathbf{R}_k$ such that

$$\mathbf{S}_k = \mathbf{R}_k \cdots \mathbf{R}_2 \mathbf{R}_1 \mathbf{AC}_1 \mathbf{C}_2 \cdots \mathbf{C}_k \geq 0 \quad \text{where} \quad \mathbf{S}_k e = e \quad \text{(all row sums are one).}$$

A square matrix $\mathbf{S}$ having nonnegative entries and row sums equal to one is called a *stochastic matrix*. When the row sums as well as the column sums in $\mathbf{S}$ are all equal to one, $\mathbf{S}$ is said to be a *doubly stochastic matrix*.

Thus the matrix $\mathbf{S}_k = \mathbf{R}_k \cdots \mathbf{R}_2 \mathbf{R}_1 \mathbf{AC}_1 \mathbf{C}_2 \cdots \mathbf{C}_k$ resulting from successive column and row scalings is a stochastic matrix for every $k = 1, 2, 3, \ldots$. The hope is that $\mathbf{S} = \lim_{k \to \infty} \mathbf{S}_k$ exists and $\mathbf{S}$ is doubly stochastic. Alas, this does not always happen, but all is not lost because there is a theorem that tells us exactly when it will happen. Understanding the theorem requires knowledge of the "diagonals" in a square matrix.

Diagonals

The main diagonal of a square matrix $\mathbf{A}$ is the set of entries $\{a_{11}, a_{22}, \ldots a_{mm}\}$, and this is familiar to most people. However, there are other "diagonals" in $\mathbf{A}$ that receive less notoriety. In general, if $\sigma = (\sigma_1, \sigma_2, \ldots, \sigma_m)$ is a permutation of the integers $(1, 2, \ldots, m)$, then the *diagonal of* **A** *associated with* σ is defined to be the set

$$\{a_{1\sigma_1}, a_{2\sigma_2}, \ldots a_{m\sigma_m}\}.$$

For example, in $\mathbf{A} = \begin{pmatrix} 1 & 2 & 3 \\ 4 & 5 & 6 \\ 7 & 8 & 9 \end{pmatrix}$ the main diagonal $\{1, 5, 9\}$ corresponds to the natural order $(1, 2, 3)$, while this and all other diagonals are derived from the six permutations as shown below.

Permutation	Diagonal
$(1, 2, 3)$	$\{a_{11}, a_{22}, a_{33}\}$
$(1, 3, 2)$	$\{a_{11}, a_{23}, a_{32}\}$
$(2, 1, 3)$	$\{a_{12}, a_{21}, a_{33}\}$
$(2, 3, 1)$	$\{a_{12}, a_{23}, a_{31}\}$
$(3, 1, 2)$	$\{a_{13}, a_{21}, a_{32}\}$
$(3, 2, 1)$	$\{a_{13}, a_{22}, a_{31}\}$

Sinkhorn–Knopp

In 1967 Richard Sinkhorn and Paul Knopp [70] proved a theorem that perfectly applies to the OD theory. Their theorem is as follows.

Sinkhorn–Knopp Theorem

Let $\mathbf{A}_{m \times m}$ be a nonnegative matrix.

- The iterative process of alternately scaling the columns and rows of **A** will converge to a doubly stochastic limit **S** if and only if there is at least one diagonal in **A** that is strictly positive, in which case **A** is said to have *support*.

This statement guarantees the scaling process will converge to a doubly stochastic matrix, but it does not ensure that the products $\mathcal{C}_k = \mathbf{C}_1 \mathbf{C}_2 \cdots \mathbf{C}_k$ and $\mathcal{R}_k = \mathbf{R}_k \cdots \mathbf{R}_2 \mathbf{R}_1$ will converge—e.g., successively scaling $\mathbf{A} = \begin{pmatrix} 1 & 1 \\ 0 & 1 \end{pmatrix}$ converges to **I** (the identity), but $\mathcal{C}_k$ and $\mathcal{R}_k$ do not converge.

- The limits of the diagonal matrices

$$\lim_{k \to \infty} \mathcal{C}_k = \mathbf{C} \quad \text{and} \quad \lim_{k \to \infty} \mathcal{R}_k = \mathbf{R} \tag{7.19}$$

exist and $\mathbf{RAC} = \mathbf{S}$ is doubly stochastic if and only if every positive entry in **A** lies on a positive diagonal, in which case **A** is said to have *total support*.

OD Matrices

As discussed on page 83, statistics other than game scores (such as yardage in football) can be used to compute OD ratings. Throughout the rest of the OD development, it will be understood that $\mathbf{A}_{m \times m}$ is a matrix of nonnegative statistics (scores, yardage, etc.), and such a matrix will be called an *OD matrix*. The major observation is that applying the Sinkhorn–Knopp process to an OD matrix produces OD ratings by reciprocation. This is developed in the following theorem.

The OD Ratings and Sinkhorn–Knopp

In subsequent discussions it is convenient to identify column vectors $\mathbf{x}$ with associated diagonal matrices by means of the mapping

$$\mathbf{x} = \begin{pmatrix} x_1 \\ x_2 \\ \vdots \\ x_m \end{pmatrix} \Longleftrightarrow \operatorname{diag}\{\mathbf{x}\} = \begin{pmatrix} x_1 & 0 & \cdots & 0 \\ 0 & x_2 & \cdots & 0 \\ \vdots & \vdots & \ddots & \vdots \\ 0 & 0 & \cdots & x_m \end{pmatrix}.$$

Notice that $\operatorname{diag}\{\mathbf{x}\}\,\mathbf{e} = \mathbf{x} = \mathbf{e}^T \operatorname{diag}\{\mathbf{x}\}$. Furthermore, for vectors $\mathbf{v}$ with nonzero components and for diagonal matrices $\mathbf{D}$ with nonzero diagonal elements, it is straightforward to verify that

$$\operatorname{diag}\{(\mathbf{Dv})^{\div}\} = \mathbf{D}^{-1}\operatorname{diag}\{\mathbf{v}^{\div}\} \quad \text{and} \quad \operatorname{diag}\{(\mathbf{Dv}^{\div})^{\div}\} = \mathbf{D}^{-1}\operatorname{diag}\{\mathbf{v}\}. \quad (7.20)$$

We are now in a position to prove the primary OD theorem that tells us when the OD rating vectors exist and when they do not.

The OD Theorem

For an OD matrix $\mathbf{A}_{m \times m}$, the OD sequences

$$\mathbf{o}_k = \mathbf{A}^T \mathbf{d}_{k-1}^{\div} \quad \text{and} \quad \mathbf{d}_k = \mathbf{A}\mathbf{o}_k^{\div}, \quad \text{with } \mathbf{d}_0 = \mathbf{e}, \quad (7.21)$$

respectively converge to OD rating vectors $\mathbf{o}$ and $\mathbf{d}$ if and only if every positive entry in $\mathbf{A}$ lies on a positive diagonal—i.e., if and only if $\mathbf{A}$ has total support.

Proof. As in earlier discussions, let $\mathcal{C}_k$ and $\mathcal{R}_k$ denote the respective products

$$\mathcal{C}_k = \mathbf{C}_1 \mathbf{C}_2 \cdots \mathbf{C}_k \quad \text{and} \quad \mathcal{R}_k = \mathbf{R}_k \cdots \mathbf{R}_2 \mathbf{R}_1$$

of the diagonal scaling matrices generated by the Sinkhorn–Knopp process. The key is to establish a connection between these products and the OD sequences in (7.21). To do this, recall that the diagonal scaling matrices in the Sinkhorn–Knopp process are generated as follows.

$$\begin{aligned}
\mathbf{C}_1 &= \operatorname{diag}\{(\mathbf{A}^T\mathbf{e})^{\div}\} & \mathbf{R}_1 &= \operatorname{diag}\{(\mathbf{A}\mathbf{C}_1\mathbf{e})^{\div}\} \\
\mathbf{C}_2 &= \operatorname{diag}\{((\mathbf{R}_1\mathbf{A}\mathbf{C}_1)^T\mathbf{e})^{\div}\} & \mathbf{R}_2 &= \operatorname{diag}\{(\mathbf{R}_1\mathbf{A}\mathbf{C}_1\mathbf{C}_2\mathbf{e})^{\div}\} \\
\mathbf{C}_3 &= \operatorname{diag}\{((\mathbf{R}_2\mathbf{R}_1\mathbf{A}\mathbf{C}_1\mathbf{C}_2)^T\mathbf{e})^{\div}\} & \mathbf{R}_3 &= \operatorname{diag}\{(\mathbf{R}_2\mathbf{R}_1\mathbf{A}\mathbf{C}_1\mathbf{C}_2\mathbf{C}_3\mathbf{e})^{\div}\} \quad (7.22) \\
&\;\;\vdots & &\;\;\vdots
\end{aligned}$$

Observe that the products $\mathcal{C}_k$ and $\mathcal{R}_k$ can be expressed as

$$\mathcal{C}_k = \operatorname{diag}\{\mathbf{c}_k\} \quad \text{and} \quad \mathcal{R}_k = \operatorname{diag}\{\mathbf{r}_k\} \quad \text{for } k = 1, 2, 3, \ldots, \quad (7.23)$$

where $r_0 = e$, and

$$c_k = (A^T r_{k-1})^{\div} \quad \text{and} \quad r_k = (Ac_k)^{\div} \quad \text{for } k = 1, 2, 3, \ldots . \tag{7.24}$$

This can be verified by writing out the first few iterates. For example, $c_1 = (A^T e)^{\div}$, so it is clear from (7.22) that $C_1 = \text{diag}\{c_1\}$, and $C_1 e = c_1$, and thus

$$R_1 = \text{diag}\{(AC_1 e)^{\div}\} = \text{diag}\{(Ac_1)^{\div}\} = \text{diag}\{r_1\}.$$

Use this along with (7.20) to write

$$C_2 = \text{diag}\{((R_1 AC_1)^T e)^{\div}\} = \text{diag}\{(C_1 A^T R_1 e)^{\div}\} = \text{diag}\{(C_1 A^T r_1)^{\div}\}$$
$$= \text{diag}\{(C_1 c_2^{\div})^{\div}\} = C_1^{-1} \text{diag}\{c_2\}$$

so that $\mathcal{C}_2 = C_1 C_2 = \text{diag}\{c_2\}$. Similarly, this together with (7.20) yields

$$R_2 = \text{diag}\{(R_1 AC_1 C_2 e)^{\div}\} = \text{diag}\{(R_1 Ac_2)^{\div}\}$$
$$= \text{diag}\{(R_1 r_2^{\div})^{\div}\} = R_1^{-1} \text{diag}\{r_2\}$$

so that $\mathcal{R}_2 = R_2 R_1 = \text{diag}\{r_2\}$. Proceeding inductively establishes (7.23). Now observe that the sequences in (7.24) are simply the reciprocals of the OD sequences in (7.21). In other words,

$$c_k = o_k^{\div} \quad \text{and} \quad r_k = d_k^{\div}. \tag{7.25}$$

Again, this is easily seen by writing out the first few terms as shown below and proceeding inductively.

$$c_1 = (A^T e)^{\div} \quad \text{and} \quad o_1 = A^T e^{\div} = A^T e \quad \Longrightarrow \quad o_1 = c_1^{\div}$$
$$r_1 = (Ac_1)^{\div} \quad \text{and} \quad d_1 = Ao_1^{\div} = Ac_1 \quad \Longrightarrow \quad d_1 = r_1^{\div}$$
$$c_2 = (A^T r_1)^{\div} \quad \text{and} \quad o_2 = A^T d_1^{\div} = A^T r_1 \quad \Longrightarrow \quad o_2 = c_2^{\div}$$
$$r_2 = (Ac_2)^{\div} \quad \text{and} \quad d_2 = Ao_2^{\div} = Ac_2 \quad \Longrightarrow \quad d_2 = r_2^{\div}$$
$$\vdots \qquad\qquad\qquad \vdots \qquad\qquad\qquad \vdots$$

The Sinkhorn–Knopp Theorem on page 89 says that the limits

$$\lim_{k \to \infty} \mathcal{C}_k = C \quad \text{and} \quad \lim_{k \to \infty} \mathcal{R}_k = R$$

exist if and only if A has total support. Consequently, (7.23) means that the limits

$$\lim_{k \to \infty} c_k = \lim_{k \to \infty} \mathcal{C}_k e = Ce \quad \text{and} \quad \lim_{k \to \infty} r_k = \lim_{k \to \infty} \mathcal{R}_k e = Re$$

exist if and only if A has total support, and thus (7.25) ensures that the limits

$$\lim_{k \to \infty} o_k = \lim_{k \to \infty} c_k^{\div} = C^{-1} e \quad \text{and} \quad \lim_{k \to \infty} d_k = \lim_{k \to \infty} r_k^{\div} = R^{-1} e$$

exist if and only if A has total support. ∎

Cheating a Bit

Not all OD matrices will have the total support property—i.e., not every positive entry will lie on a positive diagonal. Without total support, the OD rating vectors cannot exist because the defining sequences will not converge, so it is only natural to try to cheat a bit in order to force convergence. One rather obvious way of cheating a little is to simply add a small number $\epsilon > 0$ to each entry in $\mathbf{A}$. That is, make the replacement

$$\mathbf{A} \longleftarrow \mathbf{A} + \epsilon \mathbf{e}\mathbf{e}^T > 0.$$

This forces all entries to be positive, so total support is automatic, and thus convergence of the OD sequences is guaranteed. If it bothers you to add something to the main diagonal of $\mathbf{A}$ because it suggests that a team has beaten itself, you can instead make the replacement

$$\mathbf{A} \longleftarrow \mathbf{A} + \epsilon(\mathbf{e}\mathbf{e}^T - \mathbf{I}),$$

which will also guarantee total support.

If ϵ is small relative to the nonzero entries in the original OD matrix, then the spirit of the original data will not be overly compromised, and the OD ratings produced by the slightly modified matrix will still do a good job of reflecting relative offensive and defensive strength. However, there is no free lunch in the world, and there is a price to be paid for this bit of deception.

While relative offensive and defensive strengths are not overly distorted by a small perturbation, the rate of convergence can be quite slow. It should be intuitive that as ϵ becomes smaller, the number of iterations needed to produce viable results grows larger. In other words, you need to walk a tightrope by balancing the desired quality of your OD ratings against the time you are willing to spend to generate them. However, keep in mind that you may not need many significant digits of accuracy to produce useable ratings.

As demonstrated in (7.25), the OD sequences are nothing more (or less) than the reciprocals of the Sinkhorn–Knopp sequences in (7.24), so the rate of convergence of the OD sequences is the same as convergence rates of the Sinkhorn–Knopp procedure. While estimates of convergence rates of the Sinkhorn–Knopp have been established, they are less than straightforward. Some require you to know the end result of the process so that you cannot compute an a priori estimate while others are "order estimates" where the order constants are not known or not computable. A detailed discussion of convergence characteristics would take us too far astray and would add little to the application of the OD method. Technical convergence discussions are in [30, 46, 72, 73]. Additional information and applications involving the OD method can be found in Govan's work [34, 35].

ASIDE: HITS

Our OD method is a nonlinear cousin of Jon Kleinberg's algorithm called HITS (Hypertext Induced Topic Search) [45] that was designed for ranking linked documents. HITS became a pivotal technique for ranking the most popular type of linked documents, webpages. The HITS idea was incorporated into the search engine Ask.com. HITS, like many search engine models, views the Web as a gigantic directed graph. As observed in several places in this book, the interaction between competitive contests can also be viewed as a large graph. Consequently, some webpage ranking algorithms can be adapted to create rankings (and vice versa). A detailed discussion of HITS for ranking webpages is given in [45] and [49], but here is

J. Kleinberg

the basic idea in a nutshell. Each webpage, which in Figure 7.1 is represented by a node, is given both a "hub rating" and an "authority rating." A hub rating is meant to reflect the quality and quantity of a Web page's outlinks, while the authority rating measures the quality and quantity of its inlinks. The intuition is that hub pages act as central points in the Web

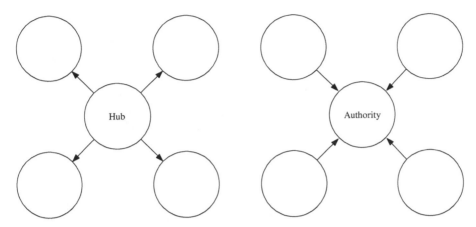

Figure 7.1 Hub page and an authority page

that provide users with many good links to follow—e.g., YAHOO! has many good hub pages. Authorities are pages that are generally informative and many good sites point to them—e.g., WIKIPEDIA has many good authority pages. A page's hub rating depends on the authority ratings of the pages it links to. In turn, a page's authority rating depends on the hub rating of those pages linking to it. Kleinberg called this a "mutually reinforcing" rating system because one rating is dependent on the other—do you notice the similarity with OD? To compute hub and authority ratings, start with a square adjacency matrix $\mathbf{L}$, where

$$\mathbf{L}_{ij} = \begin{cases} 1 & \text{if there exists an edge from node } i \text{ to node } j, \\ 0 & \text{otherwise.} \end{cases}$$

Choose an arbitrary initial authority rating vector (the uniform vector $\mathbf{a}_0 = \mathbf{e}$ is often used) and successively refine hub and authority ratings by setting

$$\mathbf{a}_{k+1} = \mathbf{L}^T \mathbf{h}_k, \quad \text{and} \quad \mathbf{h}_k - \mathbf{L}\mathbf{a}_k \tag{7.26}$$

These equations make the dependency between authority and hub ratings clear. This interdependence is broken by exactly the same "divorce" strategy that was described on page 81. That is, substitute the expression for $\mathbf{h}_k$ into the equation defining $\mathbf{a}_{k+1}$ to get

$$\mathbf{a}_{k+1} = \mathbf{L}^T \mathbf{L} \mathbf{a}_k \quad \text{and} \quad \mathbf{h}_{k+1} = \mathbf{L}\mathbf{L}^T \mathbf{h}_k.$$

Provided that they converge, the limits of these two sequences provide the desired hub and authority ratings. Readers interested in more details concerning implementation, code, and extensions of HITS should consult Chapter 11 in [49].

ASIDE: The Birth Of OD

A. Govan

During our work on the "Google book" [49] concerning the PageRank algorithm, we often mused over the possibilities of applying Web-page rating and ranking ideas to sports. After all, search-engine ranking methods are themselves just adaptations of basic principles from other disciplines (e.g., bibliometrics), so, instead of ranking Web pages, why couldn't the same ideas be applied to sports applications? Of course, they can. The Markovian methods discussed in Chapter 6 on page 67 are straightforward adaptions of the PageRank idea. But what about other Web-page ranking techniques? One afternoon in 2005 we were kicking these thoughts around, and Kleinberg's HITS came up. As described above, HITS is an elegantly simple idea, and although it doesn't scale up well, it is nevertheless effective for small collections—perhaps even superior to PageRank. It seemed natural to reinterpret Kleinberg's "hubs-authorities" concept as an "offense-defense" scheme. However it soon became apparent that a direct translation from hubs-and-authorities to offense-and-defense was not workable. But the underlying idea was too pretty to give up on, so the wheels continued to spin. It did not take long to realize that while hubs and authorities are linearly related as described in (7.26), the relationship between offense and defense is inherently nonlinear, and some sort of reciprocal relationship had to be formulated. After that, the logic flowed smoothly, and thus the OD method was born.

Shortly thereafter Anjela Govan arrived at NC State as a graduate student to study Markovian methods in data analytics, but in the course of casual conversation she became intrigued by rating and ranking. In spite of advice concerning the dubiousness of pursuing a thesis motivated by football, she followed her heart and went on to develop the connections between the OD method and the Sinkhorn–Knopp theory, which, as described in the discussion beginning on on page 87, provides the theoretical foundation for analyzing the mathematical properties of the OD method. In addition, she was instrumental in developing and analyzing several of the Markovian methods discussed in this book as well as several that were not elaborated here. And her adept programming ability produced substantial and innovative webscraping software to automatically harvest data from diverse web sites. This alone was a significant achievement because the countless number of numerical experiments that were subsequently conducted would have been more difficult without the use of her tools.

ASIDE: Individual Awards in Sports

At the end of the season, many sports crown players with individual awards. For instance, college football selects a winner for its prestigious Heisman trophy. Collegiate and professional basketball select all-offensive and all-defensive teams. Such selections are generally made by compiling and combining votes from coaches, sports writers, and sometimes even fans. These selections could easily be supplemented by the addition of scientific ranking methods such as those found in this book. However, it is not a trivial issue to select the most appropriate ranking method for a particular award. Some careful thought and clever mathematical modeling is required to achieve the appropriate aim of the award. For instance, the Markov model in Chapter 6 is a good choice for selecting an all-defensive team, whereas the OD method of this chapter might be most appropriate for the Player of the Year award, since OD incorporates both offensive and defensive ability. There are many parameters for tweaking and tailoring the models, making this an exciting avenue of future work.

ASIDE: More Movie Rankings: OD Meets Netflix

The aside "Movie Rankings: Colley and Massey meet Netflix" on page 25 introduced the Netflix Prize. Netflix sponsored a $1 million prize to improve its current recommendation system. The prize was claimed in September 2009 by the "BellKor" team of researchers. The earlier Netflix aside described how the Colley and Massey methods could be used to rank movies. In this aside, we describe how the OD method can be modified to do the same. In fact, since the matrix in this case is rectangular and OD gives dual rating vectors, we will get both a movie rating vector $\mathbf{m}$ and a user rating vector $\mathbf{u}$. Below we have chosen to display only the top-24 elements of the movie rating vector $\mathbf{m}$. The final user rating vector $\mathbf{u}$ and its corresponding ranking of users are of little interest to us, of course, $\mathbf{u}$ is of great interest to companies such as Netflix that build marketing strategies around such user information. The Netflix OD method begins with the thesis that *movie i is good and deserves a high rating m_i if it gets high ratings from good (i.e., discriminating) users. Similarly, user j is good and serves a high rating u_j when his or her ratings match the "true" ratings of the movies.* One mathematical translation of these linguistic statements produces the iterative procedure below, which is written in MATLAB style. There are several methods for quantifying the phrase "discriminating user" (see, for instance, some of the approaches of the Netflix prize winners) but one elementary approach tries to measure how far a user's ratings are from the true ratings.

```
u = e;
for i = 1 : maxiter
    m = A u;
    m = 5 (m-min(m)) / (max(m) -min(m)) ;
    u = 1 / ((R  -(R>0) . * (em' )).^2)e
end
```

Applying this algorithm to a subset of Netflix data, specifically the ratings of "power users,"[1] i.e., those users who have rated 1000 or more movies, results in the following top-24 list of movies. This list is considerably different than the lists on page 28 produced by the Colley and Massey methods.

Rank	Movie	Rank	Movie
1.	*Lord of the Rings III: The Return of the King*	13.	*Star Wars: Episode VI: Return of the Jedi*
2.	*Lord of the Rings II: The Two Towers*	14.	*Finding Nemo*
3.	*The Shawshank Redemption*	15.	*The Sixth Sense*
4.	*Lord of the Rings I: The Fellowship of the King*	16.	*The Silence of the Lambs*
5.	*Raiders of the Lost Ark*	17.	*Million Dollar Baby*
6.	*Star Wars: Episode V: The Empire Strikes Back*	18.	*The Incredibles*
7.	*The Godfather*	19.	*Hotel Rwanda*
8.	*Star Wars: Episode IV: A New Hope*	20.	*Indiana Jones and the Last Crusade*
9.	*Lost: Season 1*	21.	*Braveheart*
10.	*Schindler's List*	22.	*Saving Private Ryan*
11.	*The Usual Suspects*	23.	*The Sopranos: Season 1*
12.	*Band of Brothers*	24.	*The Godfather, Part II*

Top 24 movies ranked by Netflix OD method

[1]It was surprising for us to learn just how many power users there are in the Netflix dataset. There are over 13,000 users who have rated more than a thousand movies.

By The Numbers —

$3, 142, 000, 000 = total amount wagered on sports in Nevada in 2010.

— The Line Makers

Chapter Eight

Ranking by Reordering Methods

The philosophy thus far in this book is depicted below in Figure 8.1. We start with input data about the items that we'd like to rank and then run an algorithm that produces a rating vector, which, in turn, produces a ranking vector. The focus has been on moving from

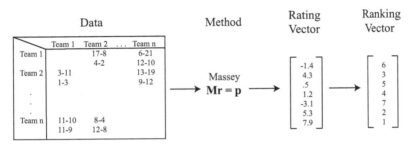

Figure 8.1 Philosophy of ranking methods of Chapters 4–7

left to right in Figure 8.1, i.e., transforming input data into a ranking vector. However, working backwards and instead turning the focus to the final product, the ranking vector itself, we are able to generate some new ideas, two of which appear in this chapter. These two new ranking methods follow the philosophy in Figure 8.2 below, which is to say, they completely bypass the rating vector.

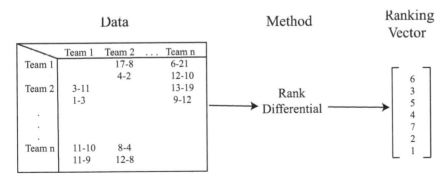

Figure 8.2 Philosophy of ranking methods of this chapter

Rank Differentials

The most apparent feature of the ranking vector is that it is a **permutation**, which, as will soon become apparent, is the keyword for this chapter. Every ranking vector of length n is a permutation of the integers 1 through n that assigns a rank position to each item. For example, for our five team example, the Massey ranking vector is

$$
\begin{array}{c}
\text{Duke} \\
\text{Miami} \\
\text{UNC} \\
\text{UVA} \\
\text{VT}
\end{array}
\begin{pmatrix}
5 \\ 1 \\ 4 \\ 3 \\ 2
\end{pmatrix},
$$

which assigns Duke to rank position 5, Miami to rank position 1, and so on. A ranking vector creates a linear ordering of the teams, yet it can also be used to consider pair-wise relationships. Pair-wise relationships are best portrayed with a matrix. Thus, the Massey ranking vector above creates the matrix of pair-wise relationships below. We call this matrix $\mathbf{R}$ since it contains information about rank differentials.

$$
\mathbf{R} =
\begin{array}{c}
\text{Duke} \\
\text{Miami} \\
\text{UNC} \\
\text{UVA} \\
\text{VT}
\end{array}
\begin{array}{ccccc}
\text{Duke} & \text{Miami} & \text{UNC} & \text{UVA} & \text{VT} \\
\end{array}
\begin{pmatrix}
0 & 0 & 0 & 0 & 0 \\
4 & 0 & 3 & 2 & 1 \\
1 & 0 & 0 & 0 & 0 \\
2 & 0 & 1 & 0 & 0 \\
3 & 0 & 2 & 1 & 0
\end{pmatrix}.
$$

The 2 in the (UVA, Duke)-entry of $\mathbf{R}$ means that UVA is two rank positions above Duke in the Massey ranking vector. Thus, the entries in the matrix correspond to dominance.

Each ranking vector of length n creates an $n \times n$ rank-differential matrix $\mathbf{R}$, which is a symmetric reordering (i.e., a row and column permutation) of the following *fundamental rank-differential matrix* $\hat{\mathbf{R}}$.

$$
\hat{\mathbf{R}}_{n \times n} =
\begin{array}{c}
1 \\ 2 \\ \vdots \\ n-1 \\ n
\end{array}
\begin{array}{ccccc}
1 & 2 & 3 & \cdots & n \\
\end{array}
\begin{pmatrix}
0 & 1 & 2 & \cdots & n-1 \\
 & 0 & 1 & \cdots & n-2 \\
 & & \ddots & \ddots & \vdots \\
 & & & \ddots & 1 \\
 & & & & 0
\end{pmatrix}.
$$

Notice the nice integer upper triangular structure of $\hat{\mathbf{R}}$, which occurs because $\hat{\mathbf{R}}$ is built from the associated fundamental ranking vector

$$
\hat{\mathbf{r}} =
\begin{array}{c}
1^{st} \\ 2^{nd} \\ 3^{rd} \\ \vdots \\ n^{th}
\end{array}
\begin{pmatrix}
1 \\ 2 \\ 3 \\ \vdots \\ n
\end{pmatrix},
$$

in which the n items appear sorted in ascending order. In summary, there is a one-to-one mapping [1] between a ranking vector and a rank-differential matrix. That is, every ranking

---

[1] The mapping is one-to-one if no ties occur in the ranking vector. See Chapter 11 for a full discussion of ties.

vector $\mathbf{r}$ creates a rank-differential matrix $\mathbf{R}$ that can be expressed as $\mathbf{R} = \mathbf{Q}^T \hat{\mathbf{R}} \mathbf{Q}$, where $\mathbf{Q}$ is a permutation matrix, and vice versa.

Because the fundamental rank-differential matrix is a team-by-team (or more generally an item-by-item) matrix, it is natural to compare it to other data matrices of the same size and type. For example, we can compare $\hat{\mathbf{R}}$ with the Markov voting matrix $\mathbf{V}_{\text{pointdiff}}$ of Chapter 6. This matrix contains the cumulative point differentials between pairs of teams. Point differentials, like rank differentials, contain information about the difference in the relative strength of two teams. However, the Markov voting matrices are created so that entries correspond to weakness. In order to relate $\mathbf{V}_{\text{pointdiff}}$ to dominance we simply compare $\mathbf{R}$ to the transpose of $\mathbf{V}_{\text{pointdiff}}$. For clarity, for the remainder of this chapter, we talk about a general data differential matrix $\mathbf{D}$. In most experiments, we use $\mathbf{D} = \mathbf{V}_{\text{pointdiff}}^T$.

We are now ready to cast the ranking problem as a *reordering* or *nearest matrix problem*. That is, given a data matrix $\mathbf{D}$, which contains information about the past pairwise comparisons of teams, find a symmetric reordering of $\mathbf{D}$ that minimizes the distance between the reordered data differential matrix and the fundamental rank-differential matrix $\hat{\mathbf{R}}$. To be mathematically precise, given $\mathbf{D}_{n \times n}$, we want to find a permutation matrix $\mathbf{Q}$ such that $\|\mathbf{Q}^T \mathbf{D} \mathbf{Q} - \hat{\mathbf{R}}\|$ is minimized. A permutation matrix $\mathbf{Q}$ applies a permutation of the integers 1 through n to the rows of the identity matrix. Thus, the optimization problem, which is defined over a permutation space, is constrained as follows.

$$\min_{\mathbf{Q}} \|\mathbf{Q}^T \mathbf{D} \mathbf{Q} - \hat{\mathbf{R}}\| \tag{8.1}$$

$$s.t. \quad \mathbf{Q}\mathbf{e} = \mathbf{e}$$
$$\mathbf{e}^T \mathbf{Q} = \mathbf{e}^T$$
$$q_{ij} \in \{0, 1\}$$

The Running Example

We use the transpose of the point differential matrix as the data differential matrix $\mathbf{D}$. Thus,

$$\mathbf{D} = \mathbf{V}^T = \begin{array}{c} \\ \text{Duke} \\ \text{Miami} \\ \text{UNC} \\ \text{UVA} \\ \text{VT} \end{array} \begin{array}{ccccc} \text{Duke} & \text{Miami} & \text{UNC} & \text{UVA} & \text{VT} \\ \left(\begin{array}{ccccc} 0 & 0 & 0 & 0 & 0 \\ 45 & 0 & 18 & 8 & 20 \\ 3 & 0 & 0 & 2 & 0 \\ 31 & 0 & 0 & 0 & 0 \\ 45 & 0 & 27 & 38 & 0 \end{array} \right) \end{array}.$$

For a 5-team example, the fundamental rank-differential matrix

$$\hat{\mathbf{R}} = \begin{array}{c} \\ 1 \\ 2 \\ 3 \\ 4 \\ 5 \end{array} \begin{array}{ccccc} 1 & 2 & 3 & 4 & 5 \\ \left(\begin{array}{ccccc} 0 & 1 & 2 & 3 & 4 \\ 0 & 0 & 1 & 2 & 3 \\ 0 & 0 & 0 & 1 & 2 \\ 0 & 0 & 0 & 0 & 1 \\ 0 & 0 & 0 & 0 & 0 \end{array} \right) \end{array}.$$

Our goal is to find the permutation of integers 1 through 5, and hence, the permutation of the five teams, that minimizes the error between the fundamental matrix $\hat{\mathbf{R}}$ and the reordered version of $\mathbf{D}$. To deal with the varying scales, we first normalize each matrix by dividing by the sum of all elements. Thus the rounded normalized versions are

$$\mathbf{D} = \begin{pmatrix} 0 & 0 & 0 & 0 & 0 \\ .19 & 0 & .08 & .03 & .08 \\ .01 & 0 & 0 & .01 & 0 \\ .13 & 0 & 0 & 0 & 0 \\ .19 & 0 & .11 & .16 & 0 \end{pmatrix} \quad \text{and} \quad \hat{\mathbf{R}} = \begin{pmatrix} 0 & 0.05 & 0.10 & 0.15 & 0.20 \\ 0 & 0 & 0.05 & 0.10 & 0.15 \\ 0 & 0 & 0 & 0.05 & 0.10 \\ 0 & 0 & 0 & 0 & 0.05 \\ 0 & 0 & 0 & 0 & 0 \end{pmatrix}.$$

Even though $n = 5$ is small, there are still $5! = 120$ different permutations to consider when reordering $\mathbf{D}$. Using brute force (i.e., total enumeration), we discover that the permutation $(5 \quad 2 \quad 4 \quad 3 \quad 1)$ is optimal, which creates the permutation matrix $\mathbf{Q}$ and, which, in turn, creates the reordered data differential matrix $\mathbf{Q}^T\mathbf{DQ}$ shown below.

$$\mathbf{Q} = \begin{pmatrix} 0 & 0 & 0 & 0 & 1 \\ 0 & 1 & 0 & 0 & 0 \\ 0 & 0 & 0 & 1 & 0 \\ 0 & 0 & 1 & 0 & 0 \\ 1 & 0 & 0 & 0 & 0 \end{pmatrix} \quad \text{and} \quad \mathbf{Q}^T\mathbf{DQ} = \begin{pmatrix} 0 & 0 & .16 & .11 & .19 \\ .08 & 0 & .03 & .01 & .16 \\ 0 & 0 & 0 & 0 & .13 \\ 0 & 0 & .01 & 0 & .01 \\ 0 & 0 & 0 & 0 & 0 \end{pmatrix}.$$

The reordered matrix $\mathbf{Q}^T\mathbf{DQ}$ is closest to the fundamental rank-differential matrix $\hat{\mathbf{R}}$. Figure 8.3 uses MATLAB's cityplot view to show a side-by-side comparison of the reordered normalized $\mathbf{D}$ and normalized $\hat{\mathbf{R}}$ matrices below.

$$\mathbf{Q}^T\mathbf{DQ} = \begin{pmatrix} 0 & 0 & .16 & .11 & .19 \\ .08 & 0 & .03 & .01 & .16 \\ 0 & 0 & 0 & 0 & .13 \\ 0 & 0 & .01 & 0 & .01 \\ 0 & 0 & 0 & 0 & 0 \end{pmatrix} \quad \text{and} \quad \hat{\mathbf{R}} = \begin{pmatrix} 0 & 0.05 & 0.10 & 0.15 & 0.20 \\ 0 & 0 & 0.05 & 0.10 & 0.15 \\ 0 & 0 & 0 & 0.05 & 0.10 \\ 0 & 0 & 0 & 0 & 0.05 \\ 0 & 0 & 0 & 0 & 0 \end{pmatrix}.$$

reordered D $\hat{R}$

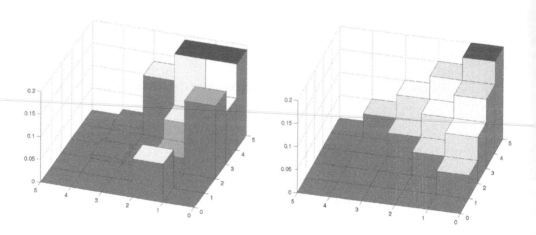

Figure 8.3 Cityplot of reordered $\mathbf{D}$ (i.e., $\mathbf{Q}\hat{\mathbf{D}}\mathbf{Q}$) and $\hat{\mathbf{R}}$ and for the 5-team example

Reordering the five teams from the original ordering of $(1 \quad 2 \quad 3 \quad 4 \quad 5)$ to the optimal ordering of $(5 \quad 2 \quad 4 \quad 3 \quad 1)$ produces a data differential matrix that is closest to the stair-step form of the fundamental rank-differential matrix $\hat{\mathbf{R}}$.

In summary, the method of rank differentials is unlike previously discussed ranking methods. In particular, it searches for the nearest rank-differential matrix to a given data matrix. And since each rank-differential matrix corresponds to a ranking, it is yet another ranking method to add to our arsenal.

The Q*bert Matrix

It's time to give credit where credit is due and in this case the credit goes to David Gottlieb, the inventor of the popular 1982 arcade game called Q*bert. Figure 8.4 shows the unmistakable similarities between the cityplot of fundamental rank-differential matrix $\hat{\mathbf{R}}$ and the Q*bert game board. In honor of Gottlieb's game, we sometimes refer to $\hat{\mathbf{R}}$, the fundamental rank-differential matrix, by the shorter and more memory-friendly name of the Q-bert matrix.

$\hat{\mathbf{R}}$ Q-bert

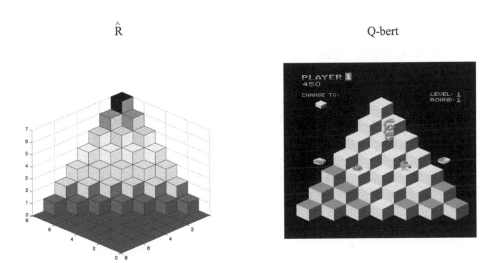

Figure 8.4 Fundamental rank-differential matrix $\hat{\mathbf{R}}$ and the Q*bert game board

Solving the Optimization Problem

Let's examine the optimization problem of Expression (8.2). The goal is to find a symmetric permutation of $\mathbf{D}$ that minimizes the difference between the permuted data matrix $\mathbf{Q}^T\mathbf{D}\mathbf{Q}$ and the fundamental rank-differential matrix $\hat{\mathbf{R}}$. For an n-item list, there are $n!$ possible permutations to consider. This, of course, means the brute force method of total enumeration that we used in the previous section is not practical. Case in point: for the tiny five-team example, to find the optimal permutation we had to check $5! = 120$ permutations. Doubling the number of teams requires consideration of $10! = 3,628,800$ permutations. Fortunately, with some clever analysis, we can avoid considering all $n!$ permutations.

The matrix formulation of the optimization problem below reveals its classification as a binary integer nonlinear program.

$$\min_{\mathbf{Q}} \|\mathbf{Q}^T \mathbf{D} \mathbf{Q} - \hat{\mathbf{R}}\|$$
$$s.t. \quad \mathbf{Q}\mathbf{e} = \mathbf{e}, \quad \mathbf{e}^T \mathbf{Q} = \mathbf{e}^T, q_{ij} \in \{0, 1\}$$

The matrix norm in the objective function affects the type of nonlinearity. For example, if the norm is of the Frobenius type, this nonlinearity is quadratic. While quadratic integer programs are challenging, there are some properties of the Frobenius norm and the Q-bert matrix $\hat{\mathbf{R}}$ that can be exploited to significantly reduce computation. For instance, with some algebra

$$\|\mathbf{Q}^T \mathbf{D} \mathbf{Q} - \hat{\mathbf{R}}\|_F^2 = trace(\mathbf{D}^T \mathbf{D}) - 2\, trace(\mathbf{Q}^T \mathbf{D} \mathbf{Q} \hat{\mathbf{R}}) + trace(\hat{\mathbf{R}}^T \hat{\mathbf{R}}).$$

Since $trace(\mathbf{D}^T \mathbf{D})$ and $trace(\hat{\mathbf{R}}^T \hat{\mathbf{R}})$ are constants, this means that the original *minimization* problem is equivalent to the *maximization* problem below.

$$\max_{\mathbf{Q}} \quad trace(\mathbf{Q}^T \mathbf{D} \mathbf{Q} \hat{\mathbf{R}})$$
$$s.t. \quad \mathbf{Q}\mathbf{e} = \mathbf{e}, \quad \mathbf{e}^T \mathbf{Q} = \mathbf{e}^T, \quad q_{ij} \in \{0, 1\}. \tag{8.2}$$

The change in the objective function brings significant computational advantages. All optimization algorithms require the computation of the objective function repeatedly and this computation is typically the most expensive step in an iterative algorithm. Fortunately, the equivalent expression of Equation (8.2) is especially efficient due to the known structure of $\hat{\mathbf{R}}$. Even more advantageous is the fact that $\hat{\mathbf{R}}$ need not be formed explicitly. In fact, the pseudocode below shows how the computation of the Frobenius objective function can be coded very efficiently.

Pseudocode to compute $trace(\mathbf{Q}^T \mathbf{D} \mathbf{Q} \hat{\mathbf{R}})$ efficiently

```
function [f]=trQtDQR(D,q);
INPUT: D = data matrix of size n × n
       q = permutation vector of size n × 1
OUTPUT: f = trace(Q^T DQR̂)
n=size(D,1);
[sortedq,index]=sort(q);
clear sortedq;
f=0;
for i=2:n
     for j=1:i-1
            f=f+(i-j)*D(index(i),index(j));
     end
end
```

The Relaxed Problem

The Frobenius formulation of the Q-bert problem gave it two tags, the quadratic tag and the integer tag. Of these two, the integer tag is more challenging. Often times, integer programs are solved approximately by the technique of relaxation, which relaxes the nasty integer constraints into continuous constraints. Thus, the relaxed Q-bert problem is given by the quadratic program below.

$$\max_{\mathbf{Q}} \quad trace(\mathbf{Q}^T \mathbf{D} \mathbf{Q} \hat{\mathbf{R}})$$
$$s.t. \quad \mathbf{Q}\mathbf{e} = \mathbf{e}, \quad \mathbf{e}^T \mathbf{Q} = \mathbf{e}^T, \quad 0 \le q_{ij} \le 1. \tag{8.3}$$

As a result, $\mathbf{Q}$ is no longer a permutation matrix, but instead is a doubly stochastic matrix (i.e., a matrix whose rows and columns sum to 1). Thus, rather than optimizing over a permutation space, we optimize over the set of doubly stochastic matrices. Sometimes, the optimal solution to the relaxed problem is integer, in which case it is also optimal for the original quadratic integer program. In this case, the extreme points of the relaxed constraint set are points in the original constraint set. Unfortunately though, the quadratic nature of the objective function allows for optimal solutions to occur in the interior of the feasible region, in which case, an optimal solution for the relaxed problem would not be feasible for the original problem. Often times the noninteger solution of the relaxed problem can be rounded to the nearest integer solution, creating a feasible solution to the original problem. Unfortunately in this case, there is no obvious rounding strategy that will guarantee that the rounded solution is anywhere near optimal for the original problem. In fact, you've really got another quadratic integer program on your hands. That is, find the nearest permutation matrix to the given doubly stochastic matrix. In summary, in our case, relaxation is disappointing.

An Evolutionary Approach

While relaxing the quadratic integer program is not promising, fortunately, another optimization approach is. Rather than trying to find a permutation matrix $\mathbf{Q}$, we can equivalently try to find a permutation vector $\mathbf{q}$ that can be used to reorder the data differential matrix $\mathbf{D}$. The reordering formulation tries to find the permutation $\mathbf{q}$ of integers 1 through n such that $\mathbf{D}(\mathbf{q}, \mathbf{q})$ is closest to $\hat{\mathbf{R}}$ in the sense that

$$\min_{\mathbf{q} \in p_n} \|\mathbf{D}(\mathbf{q}, \mathbf{q}) - \hat{\mathbf{R}}\|,$$

where $\mathbf{q} \in p_n$ means that $\mathbf{q}$ must belong to the set of all $n \times 1$ permutation vectors. For example, the permutation vector $\mathbf{q} = \begin{pmatrix} 1 \\ 3 \\ 5 \\ 4 \\ 2 \end{pmatrix}$ symmetrically reorders [2]

$$\mathbf{D} = \begin{matrix} & \begin{matrix} 1 & 2 & 3 & 4 & 5 \end{matrix} \\ \begin{matrix} 1 \\ 2 \\ 3 \\ 4 \\ 5 \end{matrix} & \begin{pmatrix} 0 & 0 & 0 & 0 & 0 \\ .19 & 0 & .08 & .03 & .08 \\ .01 & 0 & 0 & .01 & 0 \\ .13 & 0 & 0 & 0 & 0 \\ .19 & 0 & .11 & .16 & 0 \end{pmatrix} \end{matrix} \quad \text{as} \quad \mathbf{D}(\mathbf{q}, \mathbf{q}) = \begin{matrix} & \begin{matrix} 1 & 3 & 5 & 4 & 2 \end{matrix} \\ \begin{matrix} 1 \\ 3 \\ 5 \\ 4 \\ 2 \end{matrix} & \begin{pmatrix} 0 & 0 & 0 & 0 & 0 \\ .01 & 0 & 0 & .01 & 0 \\ .19 & .11 & 0 & .16 & 0 \\ .13 & 0 & 0 & 0 & 0 \\ .19 & .08 & .08 & .03 & 0 \end{pmatrix} \end{matrix}.$$

---

[2] A symmetric reordering uses the same permutation to reorder both the row and the column indices.

Notice that the elements of $\mathbf{D}$ are simply relocated in $\mathbf{D}(\mathbf{q}, \mathbf{q})$.

Matrix reorderings have been well-studied and have a rich history of uses from pre-conditioners for linear systems to preprocessors for clustering algorithms [9, 66, 7]. Thinking of the problem as a matrix reordering problem does not reduce the search space. There are still $n!$ possible permutations that reorder $\mathbf{D}$, yet optimization problems involving permutations are often tractable through a technique known as evolutionary optimization.

Evolutionary optimization, as the name suggests, takes its modus operandi from natural evolution. The idea is to start with some initial population of p possible solutions, which in this case, are permutations of the integers 1 through n. Each member of the population is evaluated for its fitness. In this case, the fitness of a solution $\mathbf{x}$ is $\|\mathbf{D}(\mathbf{x}, \mathbf{x}) - \hat{\mathbf{R}}\|$. The fittest members of the population are mated to create children that contain the best properties of their parents. Continuing with the evolutionary analogies, the less fit members are mutated in asexual fashion while the least fit members are dropped and replaced with immigrants. This new population of p permutations is evaluated for its fitness and the process continues. As the iterations proceed, it is fascinating to watch Darwin's principle of survival of the fittest. The populations march toward more evolved, fitter collections of permutations. Perhaps more fascinating is the fact that there are theorems proving that evolutionary algorithms converge to optimal solutions in many cases and near optimal solutions under certain conditions [55]. Unfortunately, evolutionary algorithms are often slow-converging, which explains their application to problems that are known as either very challenging or nearly intractable. For instance, evolutionary algorithms gained notoriety for their ability to solve larger and larger traveling salesman problems, the famously intractable problem described in the Aside on p. 104.

Figure 8.5 below is a pictorial representation of the main ideas behind evolutionary algorithms, particularly when applied to the ranking problem. This wonderful global view of the algorithm is thanks to College of Charleston graduate Kathryn Pedings. Kathryn used this figure in a few talks and as part of a poster [18]. She is a lively and interesting speaker, so it is unfortunate that readers can't hear Kathryn describing her work in person. As a paltry substitute, we have attempted to simulate Kathryn's explanation of Figure 8.5 by attaching text labels to her original figure. Simply follow the annotations in numerical order.

ASIDE: The Traveling Salesman Problem

The Traveling Salesman Problem (TSP) refers to the problem of planning a tour for a salesman who needs to visit n cities, once each. Certain variants of the problem require the salesman to start and end at the same city, presumably his hometown. The goal is to choose the minimum cost route where cost can be defined in various ways, such as total time or distance traveled. Each possible route can be considered a permutation of the n cities. Advanced variants of the TSP add constraints. For example, the salesman cannot travel more than 500 miles in a day, visit two consecutive cities in the same state, etc. Despite the elegantly simple statement of the problem, the TSP is famously hard (in fact, NP-hard, for those familiar with the term) to solve due to something called state space explosion. In its standard unconstrained formulation, the TSP has $n!/2$ possible routes that must be considered. The function $n!$ explodes quickly even when n grows slowly. See Table 8.1. In the pre-computer era, heuristic algorithms were implemented to approximate the optimal solution. A TSP with $n = 10$ cities was considered large in those days. With computers, research on TSPs took off and, in turn, so did their use.

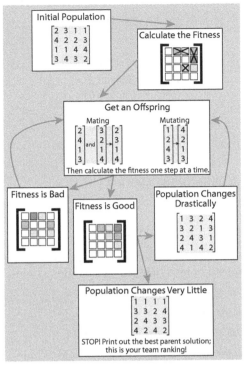

1

Each column is one member of the population. Notice that each member is a permutation vector.

2

The fitness of each member of the population is computed by reordering the matrix and counting the number of hillside violations of the reordered matrix.

3

The fittest members of the population are mated while the least fit are mutated. Mating algorithms mix properties of the parents to (hopefully) create even fitter offspring. Mutation algorithms randomly change an unfit population member.

4

The fitness of each new member of the population is calculated. The figures show how both fit and unfit members look according to hillside form.

5

The evolutionary algorithm stops when the population changes little, meaning that the randomness of the mutation operation does not find fitter members. We have a local optimum.

Figure 8.5 Overview of steps of evolutionary algorithm for ranking problem

Recently, evolutionary algorithms have led the way in TSP research. The size of TSPs that can be solved are now on the order of tens of thousands of cities. In fact, in 2006 a TSP with $n = 85,900$ cities was solved. As a result of the increase in the size of solvable TSPs, they have seen more widespread use. Today airlines use TSPs to schedule flight routes and computer hardware companies use TSPs to place copper wiring routes on circuit boards.

Advanced Rank-Differential Models

Most of the ranking methods that have been presented so far in this book allow for consideration of multiple statistics in their creation of a ranking. So too does the rank-differential method. The k multiple statistics can be combined into a single data matrix $\mathbf{D} = \alpha_1 \mathbf{D}_1 + \alpha_2 \mathbf{D}_2 + \cdots + \alpha_k \mathbf{D}_k$ with each differential matrix $\mathbf{D}_i$ given a scalar weight α_i. Then the same optimization problem, $\min_{\mathbf{q} \in p_n} \|\mathbf{D}(\mathbf{q}, \mathbf{q}) - \hat{\mathbf{R}}\|$ is solved. Or the k differential matrices can be treated separately in the optimization as

$$\min_{\mathbf{q} \in p_n} \quad \alpha_1 \|\mathbf{D}_1(\mathbf{q}, \mathbf{q}) - \hat{\mathbf{R}}\| + \alpha_2 \|\mathbf{D}_2(\mathbf{q}, \mathbf{q}) - \hat{\mathbf{R}}\| + \cdots + \alpha_k \|\mathbf{D}_k(\mathbf{q}, \mathbf{q}) - \hat{\mathbf{R}}\|.$$

Again, we recommend using an evolutionary algorithm to solve the optimization problem. Notice that the second formulation is much more expensive since calculating the fitness of each population member at each iteration requires k matrix norm computations.

Table 8.1 Explosion of $n!$ function

n	$n!$
5	120
10	$\approx 3.6 \times 10^6$
25	$\approx 1.6 \times 10^{25}$
50	$\approx 3.0 \times 10^{64}$
100	$\approx 9.3 \times 10^{157}$

Summary of the Rank-Differential Method

The Notation

$\mathbf{D}$ is the item-by-item data matrix containing pair-wise relationships such as differentials of points, yardage or other statistics.

$\hat{\mathbf{R}}$ is the fundamental rank-differential matrix.

$\mathbf{Q}$ is a permutation matrix.

$\mathbf{q}$ is the permutation vector corresponding to $\mathbf{Q}$.

$\mathbf{D}(\mathbf{q}, \mathbf{q})$ is the reordered data-differential matrix $\mathbf{Q}^T \mathbf{D} \mathbf{Q}$.

The Algorithm

1. Solve the optimization problem

$$\min_{\mathbf{q} \in p_n} \|\mathbf{D}(\mathbf{q}, \mathbf{q}) - \hat{\mathbf{R}}\|,$$

where p_n is the set of all possible $n \times 1$ permutation vectors. We recommend using an evolutionary algorithm.

2. Use the optimal solution $\mathbf{q}$ of the above problem to find the ranking vector by sorting $\mathbf{q}$ in ascending order and storing the sorted indices as the ranking vector.

Properties of the Rank-Differential Method

- All the ranking methods discussed thus far produce both a rating and a ranking vector. The rank-differential method is the first to produce just a ranking vector.

- This method finds a reordering of the data differential matrix $\mathbf{D}$ that is closest to the fundamental rank-differential matrix $\hat{\mathbf{R}}$. While closeness can be measured in many ways, it was demonstrated on page 102 that the Frobenius measure is particularly efficient.

- Because the optimization problem is very challenging (in fact, NP-hard), the rank-differential method is slower and more expensive than the other ranking methods presented in this book. Thus, it may be limited to smaller collections of items. For example, ranking sports teams is within in the scope of the method, but, unless clever implementations are discovered, ranking webpages on the full Web is not.

- Any data matrix can be used as input to the rank-differential matrix. In this chapter, we recommended using the point differential matrix of the Markov method, however, any differential data could be used.

ASIDE: The Rank-Differential Method and Graph Isomorphisms

The rank-differential method is intimately related to the graph and subgraph isomorphism problems.[3] In such cases, $\mathbf{D}$ and $\hat{\mathbf{R}}$ represent the (possibly weighted) adjacency matrices of two distinct graphs. The goal is to determine if one graph can have its nodes relabeled to produce the other. If it can, then $\|\mathbf{Q}^T\mathbf{D}\mathbf{Q} - \hat{\mathbf{R}}\| = 0$ and the two graphs are said to be isomorphic. If it cannot, then $\|\mathbf{Q}^T\mathbf{D}\mathbf{Q} - \hat{\mathbf{R}}\|$ is minimized to produce a relabeling of $\mathbf{D}$'s nodes that brings $\mathbf{D}$ closest to $\hat{\mathbf{R}}$. Our formulation is different on two counts. First, our formulation is a bit more general in that $\mathbf{D}$ and $\hat{\mathbf{R}}$ do not have to represent adjacency matrices. Second, our formulation, on the other hand, is a bit more specific because $\hat{\mathbf{R}}$ has its very special structure. Graph isomorphisms have been used successfully in a variety of applications ranging from the identification of chemical compounds to design of electronic circuit boards. Progress that we make here can be applied there, emphasizing the value of the connection between our problem and the graph isomorphism problem.

Rating Differentials

The ranking methods of Chapters 4–7 each produce a rating vector that is then sorted to give a ranking vector. The rank-differential method of the previous section is the first to bypass the rating vector altogether and produce only a ranking vector. Though the name uses the word rating, the same is true of the rating-differential method presented in this section. It skips the rating vector and goes directly to the ranking vector. Nevertheless, we will see that the word rating in the method's name is appropriate.

Consider the Massey rating vector $\mathbf{r}$ for the five team example,

$$\mathbf{r} = \begin{array}{c} \text{Duke} \\ \text{Miami} \\ \text{UNC} \\ \text{UVA} \\ \text{VT} \end{array} \begin{pmatrix} -24.8 \\ 18.2 \\ -8.0 \\ -3.4 \\ 18.0 \end{pmatrix},$$

which creates the following rating-differential matrix

$$\mathbf{R} = \begin{array}{c} \text{Duke} \\ \text{Miami} \\ \text{UNC} \\ \text{UVA} \\ \text{VT} \end{array} \begin{array}{ccccc} \text{Duke} & \text{Miami} & \text{UNC} & \text{UVA} & \text{VT} \\ \begin{pmatrix} 0 & 0 & 0 & 0 & 0 \\ 43 & 0 & 26.2 & 21.6 & .2 \\ 32.8 & 0 & 0 & 0 & 0 \\ 21.4 & 0 & 4.6 & 0 & 0 \\ 42.8 & 0 & 26 & 21.4 & 0 \end{pmatrix} \end{array}.$$

---

[3] We thank Schmuel Friedland for pointing out this connection.

We've admittedly used the same notation for both differential matrices, though this should cause no confusion. The entries in the matrix indicate the type of differential matrix. A rank-differential matrix contains only integers, while a rating-differential matrix has scalars. Unlike the rank-differential setting, where there is a single fundamental rank-differential matrix, there is no single fundamental rating-differential matrix. Yet there is a fundamental structure. Consider a rating vector that is already in the correct sorted order. For example, the rating vector

$$
\mathbf{r} = \begin{array}{c} 1 \\ 2 \\ 3 \\ 4 \\ 5 \end{array}\begin{pmatrix} .9 \\ .7 \\ .3 \\ .1 \\ -.2 \end{pmatrix} \text{ produces the rating-differential matrix } \mathbf{R} = \begin{array}{c} \\ 1 \\ 2 \\ 3 \\ 4 \\ 5 \end{array}\begin{array}{ccccc} 1 & 2 & 3 & 4 & 5 \\ \begin{pmatrix} 0 & .2 & .6 & .8 & 1.1 \\ 0 & 0 & .4 & .6 & .9 \\ 0 & 0 & 0 & .2 & .5 \\ 0 & 0 & 0 & 0 & .3 \\ 0 & 0 & 0 & 0 & 0 \end{pmatrix} \end{array},
$$

which has a very characteristic structure. Every rating-differential matrix $\mathbf{R}$ can be re-ordered to have this structure.

A rating-differential matrix $\mathbf{R}$ is in *fundamental form* if

$$r_{ij} = 0, \quad \forall\, i \geq j \qquad \text{(strictly upper triangular)}$$
$$r_{ij} \leq r_{ik}, \quad \forall\, i \ni j \leq k \qquad \text{(ascending order across rows)}$$
$$r_{ij} \geq r_{kj}, \quad \forall\, j \ni i \leq k \qquad \text{(descending order down columns)}.$$

The cityplot of $\mathbf{R}(\mathbf{q}, \mathbf{q})$ resembles a hillside. See Figure 8.6. Thus, we sometimes

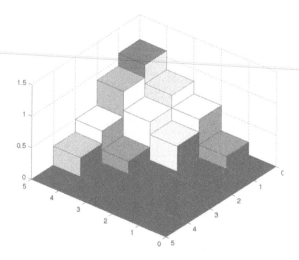

Figure 8.6 Cityplot of rating-differential matrix $\mathbf{R}$ in fundamental or hillside form

refer to the fundamental form as the *hillside form*.[4] Every rating vector creates a rating-differential matrix that can be reordered to be in fundamental form. The ordering of the items required to produce the reordering is a permutation of the integers 1 through n. And this permutation is precisely the ranking vector.

Astute readers are probably questioning us now: didn't the above logic produce a ranking vector by assuming the availability of a rating vector? How valuable is that? If we had a rating vector, then of course we have a ranking vector. The example in the next section shows how the definitions produced by this seemingly backward logic lead to a new, viable ranking method that does not require the presence of a rating vector first.

The Running Example

The rating-differential method begins with a team-by-team data matrix as input. For the same reason mentioned in the rank-differential section, it makes sense to use a matrix of differential data $\mathbf{D}$. In this example, we use the transpose of the Markov voting matrix $\mathbf{V}$ of point differentials, denoted $\mathbf{V}^T$, since here it makes sense to think of votes as an indicator of strength rather than weakness. The goal of the rating-differential method is to find a reordering of $\mathbf{D}$ that brings it to the fundamental form of the box on page 108 or as near to that form as possible. For the five team example,

$$
\mathbf{D} = \mathbf{V}^T = \begin{array}{c} \\ \text{Duke} \\ \text{Miami} \\ \text{UNC} \\ \text{UVA} \\ \text{VT} \end{array}
\begin{array}{ccccc} \text{Duke} & \text{Miami} & \text{UNC} & \text{UVA} & \text{VT} \\
\left(\begin{array}{ccccc} 0 & 0 & 0 & 0 & 0 \\ 45 & 0 & 18 & 8 & 20 \\ 3 & 0 & 0 & 2 & 0 \\ 31 & 0 & 0 & 0 & 0 \\ 45 & 0 & 27 & 38 & 0 \end{array}\right) \end{array}
$$

and there are 5! permutations of the five teams that will symmetrically reorder the matrix. Using the brute force method of total enumeration we find that the permutation

$$
\mathbf{q} = \begin{pmatrix} 2 \\ 5 \\ 3 \\ 4 \\ 1 \end{pmatrix} \implies \mathbf{D}(\mathbf{q}, \mathbf{q}) = \begin{array}{c} \\ 2 \\ 5 \\ 3 \\ 4 \\ 1 \end{array} \begin{array}{ccccc} 2 & 5 & 3 & 4 & 1 \\ \left(\begin{array}{ccccc} 0 & 20 & 18 & 8 & 45 \\ 0 & 0 & 27 & 38 & 45 \\ 0 & 0 & 0 & 2 & 3 \\ 0 & 0 & 0 & 0 & 31 \\ 0 & 0 & 0 & 0 & 0 \end{array}\right) \end{array},
$$

which, though not in fundamental form, is closest with the fewest constraint violations (just six). Sorting the optimal permutation vector $\mathbf{q}$ and keeping track of the indices produces the ranking vector

$$
\begin{array}{c} \text{Duke} \\ \text{Miami} \\ \text{UNC} \\ \text{UVA} \\ \text{VT} \end{array} \begin{pmatrix} 5 \\ 1 \\ 3 \\ 4 \\ 2 \end{pmatrix}.
$$

---

[4] Actually, we sometimes call this the R-bert form. Arcade fans might remember the 1982 game Q-bert in which a character of the same name hops around on a stair-step hillside that, with its equally spaced stairs, looks just like $\hat{\mathbf{R}}$, the fundamental rank-differential matrix of Figure 8.3. Because the rating-differential matrix has unequally spaced stairs, we jokingly call this the R-bert, rather than the Q-bert, form.

Solving the Reordering Problem

The rating-differential problem can be stated as: find the permutation $\mathbf{q}$ such that the reordered data differential matrix $\mathbf{D}(\mathbf{q}, \mathbf{q})$ violates the minimum number of constraints of the box on page 108. For a collection of n items, there are $n!$ possible permutations. The enormous state space and the presence of permutations make evolutionary optimization an attractive and appropriate methodology. Because the rank differential and rating-differential problems are so closely related, the population, mating, and mutating operations are identical. Only the fitness function must be changed for the rating-differential problem. In this case, the fitness of a permutation, which is a member of the population, is given by the number of constraints of the box on p. 108 that are violated.

While the evolutionary approach to solving the rating-differential problem can be used, the best technique is to transform the problem into a binary integer linear program (BILP) that can then be relaxed into a linear program (LP). Rather than describing this method here, we postpone that discussion until Chapter 15 as it can be shown that the rating-differential method is actually a special case of the rank-aggregation method described in that chapter. The rating-differential problem is revisited on page 194.

ASIDE: More College Basketball

The rating-differential method formed a large part of the M.S. thesis of College of Charleston graduate student Kathryn Pedings. As most graduate students do, Kathryn gave talks and posters describing her work at various conferences and universities. The most distant of these took place at Tsukuba University about an hour outside of Tokyo, Japan, where she used the following figures to illustrate her work. The figure below shows two point differential matrices for the eleven teams in the 2007 Southern Conference of men's college basketball.

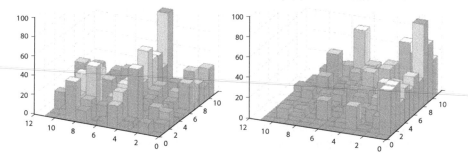

The cityplot on the left shows the matrix in its original ordering, where no structure is apparent. The cityplot on the right is the same matrix reordered according to the rating-differential method, where the hillside structure is revealed. Then in the following table she compared the rating-differential rankings (which she referred to as evolutionary optimization) with the Massey and Colley rankings.

Massey Ranking	Evolutionary Optimization	Colley Ranking
# of Viol =269	# of Viol =265	# of Viol =278

Massey	Evolutionary	Colley
Davidson	Davidson	Davidson
UNC-G	UNC-G	App. St.
GA So.	GA So.	Chatt.
App St.	App St.	GA So.
Chatt.	Chatt.	UNC-G
CofC	Elon	Elon
Elon	CofC	CofC
Wofford	W. Car.	Wofford
W. Car.	Wofford	Furman
Furman	Furman	W. Car.
Citadel	Citadel	Citadel

Notice that the rating-differential method of this section was the ranking with the fewest hillside violations. If this is a priority, then clearly the rating-differential method is the method of choice as it is guaranteed to produce a ranking with the fewest hillside violations.

Summary of the Rating-Differential Method

— *Labeling Conventions*

> D is the item-by-item data matrix containing pair-wise relationships such as differentials of points, yardage, or other statistics.
>
> q is the permutation vector that defines the reordering of D.

— *The Algorithm*

1. Solve the optimization problem

$$\min_{q \in P_n} (\text{\# of constraints listed on page 108 that } D(q, q) \text{ violates})$$

While an evolutionary algorithm works, we recommend using the Iterative LP method described later in Chapter 15.

2. The optimal solution q of the above problem is used to find the ranking vector; sort q in ascending order and store the sorted indices as the ranking vector.

— *Attributes*

- The ranking methods discussed in previous chapters of this book produce both a rating and a ranking vector. The rank-differential and the rating-differential methods both produce a ranking vector only.

- The rating-differential method finds a symmetric reordering q of the differential matrix D that brings $D(q, q)$ as close to hillside form as possible.

- Because the optimization problem is challenging (in fact, NP-hard), the rating-differential method is slower and more expensive than the other ranking methods presented in this book. Thus, it may be limited to smaller collections of items. For example, ranking sports teams is within in the scope of the method, but ranking webpages on the full Web is not. Yet despite being limited to smaller collections, the rating-differential method is a totally new approach to ranking.

- Any data matrix can be used as input to the rating-differential method. In this chapter, we recommended using a point differential matrix, however, other data differential matrices would work as well.

- Like the rank-differential method, the rating-differential method seems to work particularly well for collections of items that have many direct pair-wise relationships. That is, collections for which the item-by-item matrix has many connections. Consequently, the rating-differential method may perform poorly on a collection of items with few connections.

By The Numbers —

$18 =$ number of players that were killed in football games in 1904.

— This and the countless broken bones, fractured sculls, broken necks, wrenched legs, and many other gruesome injuries caused Theodore Roosevelt to threaten to outlaw college football in 1905.

— *The Big Scrum,* by John J. Miller

Chapter Nine

Point Spreads

Something would be amiss in a book about ratings and rankings if there was not a discussion of point spreads because beating the spread is the Holy Grail for those in the betting world. However, the goals and objectives of scientific ratings systems are not generally in line with those of bookmakers or gamblers.

A good scientific rating system tries to accurately reflect relative differences in the overall strength of each team or competitor after a reasonable amount of competition has occurred. Perhaps the most basic goal is to provide reasonable rankings that agree with expert consensus and reflect long-term ability for winning. Most of the popular scientific systems are moderately successful at producing ratings that are consistent with win percentages. But, as illustrated in the discussion of the Keener ratings on page 49, there is a big gap between estimating win percentages and estimating point spreads, and just because a system does a reasonably good job at the former is no indication of its ability to provide information concerning the latter.

An ultimate target for some is to design a system for which relative differences in the ratings that their system produces somehow leads to accurate predictions of scoring differences when two teams meet. While this is generally an impossible goal, it nevertheless continues to occupy the attention of rating system developers—after all, even Isaac Newton never lost his passion for alchemy. However, even if you achieve the rater's equivalent of turning lead into gold, it is not axiomatic that this will translate into "sure bets" because your bookmaker or casino has a different concept of "point spread."

What It Is (and Isn't)

Many novices fail to understand the true meaning of "the point spread" and the mechanisms used to construct the "spread" for a given game. The point spread put forth by bookmakers is in no way meant to be an accurate reflection or prediction of how much better one team is than another, and making accurate predictions of scoring differences is nowhere near the aim of bookies. In some sense they have an easier task. The entire purpose of the point spread that is published by a bookmaker for a given game is to attract an equal amount of money that is bet on both teams. Doing so ensures two things.

1. It helps to maximize the betting activity.

2. It balances the amount the bookie takes in against the amount he must pay out.

To understand how these two features contribute to the bookmaker's profit, you need to know exactly how a bookie makes money.

The Vig (or Juice)

The bookmaker generally does *not* make money by betting against you and having you lose. Good bookies don't care whether you win or lose your bet because their profit is usually derived from the *vig*, short for *vigorish* [1] (or *juice*, as it is sometimes called) that is simply a fee or commission that you are charged for the bookmaker's service. When the bookmaker is successful at attracting an equal amount of money that is bet on each team in a given game, then, regardless of who wins, the bookie's net pay out is zero, and his vig becomes pure profit.

Why Not Just Offer Odds?

If the bookmaker did not issue a spread but rather only took bets on which team will win a given game, then in order to maintain a balance of what is taken in against what is payed out, the payoff on a winning bet would have to be tied to the odds in favor of winning. Consequently, when a strong team is matched against a substantially weaker team, a gambler who wants to bet on the strong team is forced to risk a relatively large amount in order to win just a small amount. Furthermore, most savvy gamblers won't make a "win bet" on a likely loser even if the odds are enticing. In other words, offering only odds will usually result in decreased betting activity, and this is bad for the bookie because the total vig is directly tied to the number of bets being placed.

Somewhere around the mid 1930s Charles K. McNeil[2] realized that by offering gamblers the opportunity to bet against a published spread, bookies were able to attract more action (and thus generate more vig) than would be realized by only offering odds in favor of winning. Gambling had recently become legal in Nevada in 1931, and as soon Las Vegas bookmakers became aware of McNeil's new idea, they realized its advantage and immediately adopted it. Ever since, spread betting has been the standard mechanism for sports wagering.

How Spread Betting Works

As we have said, the bookmaker's purpose is to simultaneously maximize and balance the activity on both sides of any wager that he wants to make. He offers a point spread to create a handicap against the favored team so that instead of simply betting on the event that *team i will beat team j*, the wager is on the event that *team i will beat team j by more (or less) than the offered spread*. The bookmaker's job is to determine the point spread that will generate an equal amount of action on each side of this wager.

---

[1]This is a Yiddish slang term derived from the Russian word "vyigrysh" for winnings or profit.

[2]Charles K. McNeil (1903–1981) studied mathematics at the University of Chicago where he earned a master's degree. He went on to have at least three different careers—he taught mathematics at the Riverdale Country School, a toney prep school in New York City where John F. Kennedy was one of his students, he was a securities analyst in Chicago, and he operated his own bookmaking operation in the 1940s. His concept a of betting against a published spread changed sports wagering forever.

On the surface it may seem as though the spread should be calculated to offer each side an equal probability of winning. However, the money bet on an event is rarely equally distributed even if the wager is "fair," so spreads are revised as game time nears. Bookmaking is a black art not meant for the faint of heart, and successful bookies need to be more than numbers and stats people—they also need to be pretty good psychologists and students of human nature who have an excellent understanding of their clientele.

Beating the Spread

Just exactly what does the phrase "beating the spread" mean? Well, it can happen in two ways. If you a bet on a team and that team wins by more than the advertised point spread, then you win—you "beat the spread!" But there is another way you can do it. If you bet on the team that is not favored, and if they win, then so do you. But if they lose by fewer points than the published spread, then you still win—so again, you "beat the spread!" Let's look at a couple of examples.

> ▷ Suppose that your bookmaker has listed the BRONCOS to beat the CHARGERS by 7 points, and suppose that the score is BRONCOS = 17 and CHARGERS = 3. If you had bet on the BRONCOS, then you won because

> $$\text{BRONCOS} - \text{CHARGERS} > 7 \qquad \text{(BRONCOS covered the spread)}.$$

> If you had bet on the CHARGERS, then you lost because the CHARGERS lost by more than 7 points.

> ▷ Suppose that your bookmaker has listed the BRONCOS to beat the RAIDERS by 3 points, and suppose that the score is BRONCOS = 21 and RAIDERS = 20. If you had bet on the BRONCOS, then you lost because

> $$\text{BRONCOS} - \text{RAIDERS} < 3 \qquad \text{(BRONCOS did not cover the spread)}.$$

> However, if you had bet on the RAIDERS, then you won because they lost by less than 3 points.

To avoid a tie (also called a *push*), bookmakers frequently publish their spreads as half-point fractions such as 7.5 or 3.5. This way they avoid having to refund bets in the case of a push. In some venues, sports books are allowed to state that "ties lose" or "ties win."

Over/Under Betting

In addition to betting the spread as described above, there is a popular wager called an *over-under bet* (O/U) or a *total bet*. In this wager one does not bet on a particular team but rather on the total score in a game. The bookmaker sets a value for the "predicted" total score, and gamblers can then opt to wager that the actual total score in the game will be either over or under the bookmaker's prediction. Just as in the case of spreads, sports books will frequently specify their total to be a half-point fraction, such at 28.8, in order to avoid ties.

For example, suppose that the BRONCOS are playing the CHARGERS and the bookie sets the total at 42.5 points. If the final score is BRONCOS = 24 and CHARGERS = 16, then the total is 30, so people who made the *under* bet will win. If the final score is BRONCOS = 30, CHARGERS = 32, then the total is 62, so those who made the *over* bet will be winners. Betting the *total* is popular because it allows gamblers to exercise their sense of a game in terms of whether it will be a defensive struggle or an offensive exhibition without having to pick the actual winner.

Just as in the case of setting the spread, the bookmaker's job is *not* to actually predict the total score, but rather he needs to offer a total value that will attract the same amount of money to *over* as is attracted to *under*. Similar to spread betting, the customary vig is in the neighborhood of 4.55%.

Sports books will often allow over-under bets to be used in conjunction with spread bets. For example betting "the BRONCOS with the *over*" will be paid (somewhere in the neighborhood of 13-to-5) provided that the BRONCOS cover the spread *and* the total score is higher than the bookmaker's "prediction." While the most common over-under bet involves the total score, other game statistics such as those given below may be used.

— A football team's total rushing or passing yards, interceptions, third down conversions, etc.

— A basketball player's (or the team's) total assists, blocks, turnovers, steals, field-goal percentage, etc.

— A baseball player's (or the team's) total number of home runs, RBIs, etc.

Why Is It Difficult for Ratings to Predict Spreads?

Ok, so covering the bookmaker's spread is a different matter than predicting the actual difference in scores for a given game, but the question concerning the use of ratings to predict actual point spreads nevertheless remains. It is usually the case that even the best rating systems do less than a good job at predicting scoring differences. And among all sports, predicting football spreads from ratings is particularly troublesome. But why?

Well, there are a couple of things that limit a rating system's ability to predict spreads. The first is simply the general difference of purpose. That is, the aim of a good rating system is to provide a quantitative distillation of a team's overall strength relative to others in the same league. But in the process of distillation many of the details that actually affect scoring differences are filtered out. For example, an exciting football team may have an explosive passing game but a weak running attack and a mediocre defense. Consequently their overall rating is apt to be middling, but their games tend to be high scoring blowouts or complete busts. On the other hand a more boring football team having a balanced but otherwise mediocre offense and defense might also have a middling rating, but their score differentials will be constrained to a narrower range. In other words, a single rating number simply cannot capture all of the subtleties required to accurately predict point spreads. Estimating spreads requires separate analyses of individual aspects of a competition—you might begin by separating offensive ratings from defensive ratings as discussed in Chapter 7.

Another reason why rating differences do not adequately reflect scoring differences is because of irregular scoring patterns that are sometimes inherent in a given sport. This is particularly graphic in American football in which spreads tend to accumulate at gaps following a pattern of $3 - 4 - 3 - 4 - \cdots$. For example, the point-spread distribution for the NFL 2009–2010 season is shown in Figure 9.1.

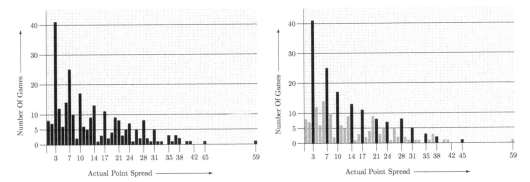

Figure 9.1 NFL 2009–2010 point-spread distribution

The graph on the right-hand side of Figure 9.1 clearly shows that the spreads in the NFL predominately spike at the $3, 7, 10, 14, 17, 21, \cdots$ positions. This is not surprising in light of the way that points are scored in football. However, these spikes in the distribution make it difficult to accurately translate rating differences into predicted scoring differences. The best that you could reasonably hope for is to have differences in your rating system somehow pick out the most probable spike—i.e., pretend that spreads other than $3, 7, 10, 14, 17, 21, \cdots$ are not possible and try to fit the distribution of the black bars on the right-hand side of Figure 9.1. But even if you are successful at this, you can still accumulate a large error over the course of an entire season.

Using Spreads to Build Ratings (to Predict Spreads?)

To further illustrate the issues involved in building a rating and ranking system that will reflect point spreads, let's pretend that we have a crystal ball that allows us to look into the future so that we already know all of the spreads for all of the games in an upcoming season. Similar to Massey's approach described on page 9, the goal is to use this information to build an optimal rating system $\{r_1, r_2, \ldots, r_n\}$ in which the differences $r_i - r_j$ come as close as possible to reflecting the point spread when team i plays team j.

To do this, first suppose that in some absolutely perfect universe there is a perfect rating vector

$$\mathbf{r} = \begin{pmatrix} r_1 \\ r_2 \\ \vdots \\ r_n \end{pmatrix}$$

such that each rating difference $r_i - r_j$ is equal to each scoring difference $S_i - S_j$, where S_i and S_j are the respective numbers of points scored by teams i and j when they meet. If the scoring differences and rating differences are respectively placed in matrices

$$\mathbf{K} = \begin{pmatrix} 0 & S_1 - S_2 & \cdots & S_1 - S_n \\ S_2 - S_1 & 0 & \cdots & S_2 - S_n \\ \vdots & \vdots & \ddots & \vdots \\ S_n - S_1 & S_n - S_2 & \cdots & 0 \end{pmatrix} \text{ and } \mathbf{R} = \begin{pmatrix} 0 & r_1 - r_2 & \cdots & r_1 - r_n \\ r_2 - r_1 & 0 & \cdots & r_2 - r_n \\ \vdots & \vdots & \ddots & \vdots \\ r_n - r_1 & r_n - r_2 & \cdots & 0 \end{pmatrix},$$

then

$$\mathbf{K} = \mathbf{R} = \mathbf{re}^T - \mathbf{er}^T, \quad \text{in which } \mathbf{e} \text{ is a column of ones.}$$

This means that in the ideal case the *score-differential matrix* $\mathbf{K}$ is a rank-two skew symmetric matrix—it is skew symmetric because $k_{ij} = -k_{ji}$ implies that $\mathbf{K}^T = -\mathbf{K}$, and the rank is two because $rank(\mathbf{re}^T - \mathbf{er}^T) = 2$ provided that $\mathbf{r}$ is not a multiple of $\mathbf{e}$ (i.e., provided $\mathbf{R} \neq \mathbf{0}$). But we don't live in a perfect universe, so, as suggested in [33], the best that we can do is to construct a ratings vector $\mathbf{r}$ that minimizes the distance between the *score-differential matrix* $\mathbf{K}$ and the *rating-differential matrix* $\mathbf{R} = \mathbf{re}^T - \mathbf{er}^T$. This is equivalent to finding a vector $\mathbf{x}$ such that

$$f(\mathbf{x}) = \|\mathbf{K} - \mathbf{R}(\mathbf{x})\|^2 = \|\mathbf{K} - (\mathbf{xe}^T - \mathbf{ex}^T)\|^2 \tag{9.1}$$

is minimal for some matrix norm. The choice of the norm is somewhat arbitrary, so let's use the simplest (and most standard) one—namely the *Frobenius norm*, which is

$$\|\mathbf{A}\|_F = \sqrt{\sum_{i,j} a_{ij}^2} = \sqrt{trace\,(\mathbf{A}^T\mathbf{A})} \qquad \text{(see [54, page 279]).}$$

Minimizing the function $f(\mathbf{x})$ in 9.1 is accomplished by using elementary calculus—i.e., look for critical points by setting $\partial f / \partial x_i = 0$ for each i and solving the resulting system of equations for $x_1, x_2, \ldots, x_n$. The trace function has the following properties.

$$trace\,(\alpha\mathbf{A} + \mathbf{B}) = \alpha\,trace\,(\mathbf{A}) + trace\,(\mathbf{B}),$$

and

$$trace\,(\mathbf{AB}) = trace\,(\mathbf{BA}) \qquad \text{(see [54, page 90]).}$$

Using these properties yields

$$f(\mathbf{x}) = trace\,[\mathbf{K} - (\mathbf{xe}^T - \mathbf{ex}^T)]^T[\mathbf{K} - (\mathbf{xe}^T - \mathbf{ex}^T)]$$

$$= trace\,\mathbf{K}^T\mathbf{K} - trace\,[\mathbf{K}^T(\mathbf{xe}^T - \mathbf{ex}^T) + (\mathbf{xe}^T - \mathbf{ex}^T)^T\mathbf{K}]$$

$$+ trace\,[(\mathbf{xe}^T - \mathbf{ex}^T)^T(\mathbf{xe}^T - \mathbf{ex}^T)],$$

so the skew symmetry of $\mathbf{K}$ implies that

$$f(\mathbf{x}) = trace\,(\mathbf{K}^T\mathbf{K}) - 4\mathbf{x}^T\mathbf{Ke} + 2n(\mathbf{x}^T\mathbf{x}) - 2(\mathbf{e}^T\mathbf{x})^2.$$

Since $\partial\mathbf{x}/\partial x_i = \mathbf{e}_i$, differentiating f with respect to x_i produces

$$\frac{\partial f}{\partial x_i} = -4\mathbf{e}_i^T\mathbf{Ke} + 4n x_i - 4\mathbf{e}^T\mathbf{x},$$

and setting $\partial f / \partial x_i = 0$ gives us

$$n x_i - \sum_{j=i}^{n} x_j = (\mathbf{Ke})_i \quad \text{for } i = 1, 2, \ldots, n.$$

This is simply a system of linear equations that can be written as the matrix equation

$$\begin{pmatrix} (n-1) & -1 & \cdots & -1 \\ -1 & (n-1) & \cdots & -1 \\ \vdots & \vdots & \ddots & \vdots \\ -1 & -1 & \cdots & (n-1) \end{pmatrix} \begin{pmatrix} x_1 \\ x_2 \\ \vdots \\ x_n \end{pmatrix} = \begin{pmatrix} (\mathbf{Ke})_1 \\ (\mathbf{Ke})_2 \\ \vdots \\ (\mathbf{Ke})_n \end{pmatrix},$$

or equivalently,

$$\left(\mathbf{I} - \frac{\mathbf{ee}^T}{n} \right) \mathbf{x} = \frac{\mathbf{Ke}}{n}. \tag{9.2}$$

Because $\mathbf{e}^T \mathbf{Ke}$ is a scalar and $\mathbf{K}^T = -\mathbf{K}$, it follows that

$$\mathbf{e}^T \mathbf{Ke} = (\mathbf{e}^T \mathbf{Ke})^T = \mathbf{e}^T \mathbf{K}^T \mathbf{e} = -\mathbf{e}^T \mathbf{Ke} \implies \mathbf{e}^T \mathbf{Ke} = 0,$$

so the vector $\mathbf{x} = \mathbf{Ke}/n$ is one solution for the system in (9.2). Moreover, it can be shown that

$$\left(\mathbf{I} - \frac{\mathbf{ee}^T}{n} \right) \mathbf{e} = \mathbf{0} \quad \text{and} \quad rank \left(\mathbf{I} - \frac{\mathbf{ee}^T}{n} \right) = n - 1,$$

so all solutions of the system (9.2) are of the form

$$\mathbf{x} = \frac{\mathbf{Ke}}{n} + \alpha \mathbf{e}, \quad \alpha \in \Re.$$

By imposing some reasonable constraint, such as forcing the ratings to sum to zero, the scalar α can be evaluated, and thus a unique solution is determined. Indeed, if we require that $0 = \sum x_i = \mathbf{e}^T \mathbf{x}$, then

$$0 = \mathbf{e}^T \mathbf{x} = \frac{\mathbf{e}^T \mathbf{Ke}}{n} + \alpha \mathbf{e}^T \mathbf{e} = \alpha \mathbf{e}^T \mathbf{e} \implies \alpha = 0,$$

and thus

$$\mathbf{r} = \frac{\mathbf{Ke}}{n} \quad \text{(which is the centroid (or mean) of the columns in } \mathbf{K})$$

is the only critical point for $f(\mathbf{x})$ that satisfies the constraint $\mathbf{e}^T \mathbf{x} = 0$. Recall that being a critical point is only a necessary condition for being a point at which a relative minimum or maximum is attained, so we still have to check to see if $\mathbf{r} = \mathbf{Ke}/n$ is indeed a minimizer. A sufficient condition for a critical point to provide a local minimum is for the Hessian [54, page 570] to be positive definite at the point, but, unfortunately, the Hessian for the case at hand is only positive semidefinite, so some additional work is required. To verify that

$$\min_{\mathbf{e}^T \mathbf{x} = 0} f(\mathbf{x}) = \min_{\mathbf{e}^T \mathbf{x} = 0} \| \mathbf{K} - (\mathbf{xe}^T - \mathbf{ex}^T) \|_F^2$$

is attained at $\mathbf{r} = \mathbf{Ke}/n$, consider points in a neighborhood of $\mathbf{r}$ that satisfy the constraint. That is, consider $\mathbf{r} + \varepsilon$, where $\varepsilon \neq \mathbf{0}$ but $\mathbf{e}^T \varepsilon = 0$, and write

$$f(\mathbf{r} + \varepsilon) - f(\mathbf{r}) = -4\varepsilon^T \mathbf{Ke} + 2n\|\mathbf{r} - \varepsilon\|_2^2 - 2n\|\mathbf{r}\|_2^2 - 2(\mathbf{e}^T \varepsilon)^2$$

$$= -4n\varepsilon^T \mathbf{r} + 2n(\|\mathbf{r}\|_2^2 + \|\varepsilon\|_2^2 + 2\varepsilon^T \mathbf{r}) - 2n\|\mathbf{r}\|_2^2 - 2(\mathbf{e}^T \varepsilon)^2$$

$$= 2 \left(n\|\varepsilon\|_2^2 - (\mathbf{e}^T \varepsilon)^2 \right). \tag{9.3}$$

The Cauchy–Schwarz (or CBS) inequality [54, page 287] says that $\mathbf{e}^T \varepsilon \leq \|\mathbf{e}\|_2 \|\varepsilon\|_2$ with equality holding if and only if $\varepsilon = \alpha \mathbf{e}$ for some α. Equality cannot hold for our problem because $\mathbf{e}^T \varepsilon = 0$ and $\varepsilon \neq 0$, so applying CBS to (9.3) leads to the conclusion that

$$f(\mathbf{r} + \varepsilon) - f(\mathbf{r}) > 0.$$

Therefore, the "best" ratings vector determined by point spreads is $\mathbf{r} = (\mathbf{Ke})/n$. Below is a summary of these observations.

The "Best" Ratings Derived from Spreads

The "best" set of ratings $\{r_1, r_2, \ldots, r_n\}$ with $\sum_i r_i = 0$ that can be determined from point spreads in the sense that the (Frobenius) distance between the score-differential matrix $\mathbf{K} = [S_i - S_j]$ and the rating-differential matrix $\mathbf{R} = [r_i - r_j]$ is minimized is given by the entries in the vector

$$\mathbf{r} = \frac{\mathbf{Ke}}{n} = \text{the centroid (or mean) of the columns in } \mathbf{K}. \qquad (9.4)$$

In the future, the entries in $\mathbf{r} = \mathbf{Ke}/n$ will be called the *spread ratings* or the *centroid ratings*. For the function $f(\mathbf{x})$ in (9.1) and for $\mathbf{R} = \mathbf{re}^T - \mathbf{er}^T$, it can be shown that

$$\min_{\mathbf{e}^T\mathbf{x}=0} f(\mathbf{x}) = f(\mathbf{r}) = \|\mathbf{K} - \mathbf{R}\|_F^2 = \|\mathbf{K}\|_F^2 - \|\mathbf{R}\|_F^2.$$

NFL 2009–2010 Spread Ratings

To illustrate the use of the spread ratings in (9.4), let's use the actual score differences that occurred in the 267 games in the 2009–2010 NFL season. If teams i and j met more than once, then matrix $\mathbf{K}$ is constructed by letting S_i and S_j be the respective average number of points that the teams score. The resulting spread ratings are shown below.

Rank	Team	Rating	Rank	Team	Rating
1.	SAINTS	6.2187	17.	PANTHERS	-0.1094
2.	VIKINGS	4.7187	18.	BRONCOS	-0.1250
3.	PACKERS	3.9687	19.	TITANS	-0.9531
4.	RAVENS	3.6875	20.	GIANTS	-1.0625
5.	PATRIOTS	3.6250	21.	REDSKINS	-1.0937
6.	JETS	3.0000	22.	DOLPHINS	-1.1094
7.	COLTS	2.8594	23.	BEARS	-1.5781
8.	CHARGERS	2.7031	24.	BILLS	-1.7812
9.	EAGLES	2.5469	25.	JAGUARS	-2.9531
10.	TEXANS	2.0781	26.	CHIEFS	-3.0156
11.	COWBOYS	2.0156	27.	BROWNS	-3.0469
12.	STEELERS	1.4219	28.	SEAHAWKS	-3.3281
13.	FALCONS	1.1719	29.	BUCS	-3.9687
14.	CARDINALS	0.3906	30.	RAIDERS	-5.1562
15.	BENGALS	0.3281	31.	LIONS	-5.4219
16.	NINERS	0.1875	32.	RAMS	-6.2187

These spread ratings correctly pick the winners in 191 out of the 267 NFL games in the 2009–2010 season for a hindsight prediction percentage of 71.54%. This is pretty good for such a simple technique—all that is required is just averaging the score differences (i.e., computing the row sums of $\mathbf{K}$ and dividing by 32).

Just how well do these spread ratings do in estimating the actual point spreads? The discussion on page 116 has already explained why we shouldn't have high expectations, but let's nevertheless have a look. The spread ratings do not take home field advantage into account, so, to be fair, we should allow home-field advantage factors to be included. Use linear regression to let the ratings determine the optimal "home-field advantage" parameters, α, β, and γ by using the linear model

$$\text{Expected } [(\text{home-team score}) - (\text{away-team score})] = \alpha + \beta r_h - \gamma r_a \qquad (9.5)$$

where r_h and r_a are the respective ratings for the home and away teams. The parameters, α, β, and γ are determined by computing the least squares solution to the system of equations

$$\alpha + \beta r_{h_i} - \gamma r_{a_i} = S_{h_i} - S_{a_i}, \quad i = 1, 2, \ldots, 267$$

where r_{h_i} and r_{a_i} are the respective ratings for the home and away teams in game i, and S_{h_i} and S_{a_i} are the respective scores made by the home and away teams in game i. In terms of matrices this boils down to computing the least squares solution for the linear system $\mathbf{Ax} = \mathbf{b}$, or, equivalently, the solution of the 3×3 system of normal equations $\mathbf{A}^T \mathbf{Ax} = \mathbf{A}^T \mathbf{b}$, where

$$\mathbf{A}_{267 \times 3} = \begin{pmatrix} 1 & r_{h_1} & r_{a_1} \\ 1 & r_{h_2} & r_{a_2} \\ \vdots & \vdots & \vdots \\ 1 & r_{h_{267}} & r_{a_{267}} \end{pmatrix}, \quad \mathbf{x} = \begin{pmatrix} \alpha \\ \beta \\ \gamma \end{pmatrix}, \quad \text{and} \quad \mathbf{b} = \begin{pmatrix} S_{h_1} - S_{a_1} \\ S_{h_2} - S_{a_2} \\ \vdots \\ S_{h_{267}} - S_{a_{267}} \end{pmatrix}.$$

Our computer says that $\alpha = 2.3671$, $\beta = 2.4229$, and $\gamma = 2.2523$, (rounded to five significant places), so estimating the point spread for each game is accomplished by setting

Estimated $[(\text{home score}) - (\text{away score})$ for game $i] = 2.3671 + 2.4229 r_{h_i} - 2.2523 r_{a_i}$.

The ***total absolute-spread error*** is the absolute deviation between the estimated score differences and the actual score differences for the entire season, and it is calculated to be

$$\sum_{i=1}^{267} \left| (2.3671 + 2.4229 r_{h_i} - 2.2523 r_{a_i}) - (S_{h_i} - S_{a_i}) \right| = 2827.6,$$

which turns out to be 10.59 points per game. To get a sense of how good or bad this is, some comparisons with other respected systems along with the actual Vegas lines might be considered.

Some Shootouts

Perhaps two of the best known and most widely respected raters are Jeff Sagarin who has been providing sports ratings for *USA Today* since 1985 and Ken Massey (Chapter 2) who is a contributor to the BCS college football rating system. Their respective final 2009–2010

NFL ratings that are shown below were taken from the following sites that were active at the time of this writing.

www.usatoday.com/sports/sagarin/nfl09.htm

www.masseyratings.com

Rank	Team	Rating	Rank	Team	Rating
1.	SAINTS	32.18	17.	BENGALS	21.07
2.	COLTS	29.86	18.	GIANTS	20.52
3.	VIKINGS	27.91	19.	BRONCOS	20.27
4.	JETS	27.69	20.	NINERS	20.21
5.	CHARGERS	25.99	21.	TITANS	19.76
6.	PATRIOTS	25.89	22.	BILLS	18.18
7.	COWBOYS	25.65	23.	BEARS	16.48
8.	RAVENS	25.58	24.	JAGUARS	16.2
9.	FALCONS	24.3	25.	BUCS	13.06
10.	PACKERS	24.19	26.	REDSKINS	12.84
11.	EAGLES	23.79	27.	BROWNS	12.66
12.	PANTHERS	22.87	28.	RAIDERS	12.56
13.	TEXANS	22.49	29.	CHIEFS	11.93
14.	DOLPHINS	21.43	30.	SEAHAWKS	11.61
15.	CARDINALS	21.37	31.	LIONS	6.41
16.	STEELERS	21.24	32.	RAMS	3.82

NFL 2009–2010 Sagarin ratings

Rank	Team	Rating	Rank	Team	Rating
1.	SAINTS	1.693	17.	TITANS	0.110
2.	COLTS	1.416	18.	DOLPHINS	0.062
3.	CHARGERS	1.291	19.	GIANTS	0.036
4.	COWBOYS	1.002	20.	NINERS	-0.093
5.	VIKINGS	0.909	21.	BRONCOS	-0.182
6.	JETS	0.897	22.	BILLS	-0.232
7.	EAGLES	0.814	23.	BEARS	-0.478
8.	PACKERS	0.783	24.	JAGUARS	-0.621
9.	FALCONS	0.671	25.	BUCS	-0.706
10.	PATRIOTS	0.667	26.	REDSKINS	-0.938
11.	PANTHERS	0.639	27.	BROWNS	-0.947
12.	RAVENS	0.602	28.	RAIDERS	-0.986
13.	TEXANS	0.340	29.	CHIEFS	-1.050
14.	CARDINALS	0.318	30.	SEAHAWKS	-1.451
15.	BENGALS	0.230	31.	LIONS	-2.385
16.	STEELERS	0.178	32.	RAMS	-2.591

NFL 2009–2010 Massey ratings

Sagarin and Massey are short on details (most raters don't want you to know all of the secret ingredients in their magic potions), but Sagarin states that his ratings represent a "synthesis" of two systems—an Elo system as described in Chapter 5 on page 53 that takes only winning and losing into account and some other unspecified system that considers scoring margins. Massey says that his ratings are computed as maximum likelihood estimates and are somehow tempered by a Bayesian correction—his description is too vague to implement. However, the reason that Sagarin and Massey have been singled out for comparison is because of their visibility and the fact that they have established a good record over the course of many seasons.

The results of the 2009–2010 NFL season are used below in (9.6) to compare hindsight win accuracy (percentage) and spread errors (average points per game) for six rating systems—the spread or centroid ratings on page 120; Sagarin ratings on page 122; Elo ratings on page 58; Keener ratings on page 44; Massey ratings on page 122; and, for the sake of curiosity, we also included the results derived from the Vegas betting lines that were published daily in the Raleigh, NC *News and Observer*. To make hindsight predictions from Sagarin's ratings, we added 2.96 for the home team (his suggestion). Hindsight predictions for the other systems were obtained from the linear model described in (9.5) on page 121, (although Massey suggests something more complicated for his system).

Hindsight Win Accuracy:	Average Spread Error (Points Per Game):	
Spread Ratings $= 71.5\%$	Spread Ratings $= 10.59$ points per game	
Sagarin Ratings $= 70.8\%$	Sagarin Ratings $= 10.53$ points per game	
Elo Ratings $\quad = 75.3\%$	Elo Ratings $\quad = 10.71$ points per game	(9.6)
Keener Ratings $= 73.4\%$	Keener Ratings $= 10.54$ points per game	
Massey Ratings $= 72.7\%$	Massey Ratings $= 10.65$ points per game	
Vegas Line $\quad = 67.8\%$	Vegas Line $\quad = 11.44$ points per game	

Comparisons of six rating systems using the 2009–2010 NFL season

As (9.6) shows, the hindsight accuracy for both wins and spreads of the five rating systems are all in the same ball park while the Vegas line lags slightly behind. A point being made here is that the spread (or centroid) ratings are competitive with the others for reflecting relative overall strength, but when we take into account that spread (or centroid) ratings are trivial to compute (you could do it on the back of an envelope), it makes them particularly attractive, and this is what sets the centroid method apart from other rating systems.

The performance of the Vegas line as reflected at the bottom of table (9.6) may look odd to you, but it's not when you consider it in the proper perspective. As explained on page 113, spreads published by bookmakers are not intended to reflect or predict actual spreads or winners. Nor do bookmakers care about making a statement about the overall relative strength of each team at the end of a season, which is contrary to other rating and ranking systems that may be involved in making final BCS rankings or seeding NCAA basketball tournaments. In other words, all bookmakers care about is looking ahead to the next game of the season, so hindsight accuracy is not a goal. The foresight accuracy of the Vegas lines are typically better than their hindsight accuracy, which is not surprising because betting lines tend to reflect the combined wisdom of a large and diverse pool of people who are focused on a single upcoming game. For example, while the hindsight win accuracy for the Elo system in (9.6) is significantly better than that of the Vegas lines, the Vegas lines beat Elo in foresight accuracy. Recall from page 62 that for the 2009–2010 NFL season the foresight win accuracy for the Vegas lines was 67.8%, whereas the foresight accuracy of the Elo system was slightly smaller at 65.9%.

There is no shortage of NFL ratings (or any other kind of ratings) available on the World Wide Web, and a Google search will reveal more than what most people will care to digest. The trouble with sports ratings that are available on the Web is that they generally do not provide enough technical details to be helpful in learning how they were created or in helping you design your own system. The information in the many articles cited in the Epilogue on page 225 also hints at the vast amount of effort that has been devoted to sports ratings, but these articles are more specific for those who crave more details. If you are interested in comparisons of rating systems other than those that we have highlighted, then look at Todd Beck's prediction tracker site

www.thepredictiontracker.com/nflresults.php

(which was active at the time of this writing). This site shows that there is remarkable little difference in the performance of the better NFL rating systems—i.e., it is hard to stand out.

Other Pair-wise Comparisons

If you understood the derivation that produced the spread (or centroid) ratings (9.4), then it may have occurred to you that the "spreads" k_{ij} in the matrix $\mathbf{K}$ on page 118 can be interpreted more broadly than simply as scoring differences. For example, setting

$$k_{ij} = \begin{cases} +1 & \text{if team } i \text{ beats team } j, \\ 0 & \text{if teams } i \text{ and } j \text{ tie}, \\ -1 & \text{if team } i \text{ beats team } j, \end{cases}$$

is a more basic pair-wise comparison of teams i and j than that provided by scoring differences, but $\mathbf{K}$ nevertheless remains skew symmetric, and exactly the same analysis holds to provide "spread ratings" $\mathbf{r} = \mathbf{K}e/n$ in a more general sense. This interpretation produces the following "win-loss" spread ratings for the 2009–2010 NFL season.

Rank	Team	Rating	Rank	Team	Rating
1.	SAINTS	0.375	17.	PANTHERS	0.03125
2.	CHARGERS	0.25	18.	TITANS	0
3.	COLTS	0.25	19.	GIANTS	-0.03125
4.	VIKINGS	0.21875	20.	JAGUARS	-0.03125
5.	EAGLES	0.15625	21.	NINERS	-0.03125
6.	PACKERS	0.15625	22.	BEARS	-0.09375
7.	COWBOYS	0.125	23.	DOLPHINS	-0.09375
8.	JETS	0.125	24.	BILLS	-0.15625
9.	RAVENS	0.125	25.	BROWNS	-0.15625
10.	STEELERS	0.125	26.	REDSKINS	-0.15625
11.	TEXANS	0.125	27.	CHIEFS	-0.21875
12.	PATRIOTS	0.09375	28.	RAIDERS	-0.21875
13.	CARDINALS	0.0625	29.	SEAHAWKS	-0.21875
14.	BENGALS	0.03125	30.	BUCS	-0.28125
15.	BRONCOS	0.03125	31.	LIONS	-0.28125
16.	FALCONS	0.03125	32.	RAMS	-0.34375

NFL 2009–2010 win-loss spread ratings

Naturally, these win-loss spread ratings are a more coarse assessment than those provided by score differences, but they nevertheless afford some information. For example, measured by the ability to win games, these ratings indicate that the CHARGERS and COLTS were on par with each other. Similarly, the EAGLES and PACKERS were indistinguishable, as were the COWBOYS, JETS, RAVENS, STEELERS, and TEXANS. And there are clearly other teams that group together in winning ability.

Depending on what you are rating or ranking, there are countless other ways to make pair-wise comparisons. For example, some situations might be better analyzed by setting $k_{ij} = S_i/S_j$, or even $k_{ij} = \log(S_i) - \log(S_j)$. You are only limited by your imagination.

Conclusion

The next time you see a point spread published in a betting line, remember that the person, casino, or whoever is taking the bet doesn't necessarily believe the favored team is that many points better than the underdog, but rather that they are betting on the expectation that their "predicted" point spread will generate an equal amount of money wagered on each team in the contest.

Perhaps the larger lesson is to not put an undue amount of faith in anyone's rating system for the purpose of predicting point spreads (or even picking winners)—you would be expecting too much from a single list of numbers. If after noting this you still want to make ball-park spread predictions, then simply averaging past scoring differences as described in (9.4) on page 120 will probably give you an estimate that is competitive with more complicated techniques.

By The Numbers —

  24 = the largest NFL spread ever offered by Las Vegas oddsmakers.

    — December, 1976, Pittsburgh over Tampa Bay.
    Pittsburgh covered the spread by beating Tampa Bay 42–0.

                              — sports.espn.go.com/nfl/news

Chapter Ten

User Preference Ratings

The "spread rating" ideas that were introduced in (9.4) on page 120 transcend sports ratings and rankings. A major issue that has arisen in recent years in the wake of online commerce is that of rating and ranking items by user preference. While they may not have been the first to employ user rating and ranking systems for product recommendation, companies such as Amazon.com and Netflix have developed highly refined (and proprietary) techniques for online marketing based on product recommendation systems. It would require another book to delve into all of the details surrounding these technologies, but we can nevertheless hint at how some of the ideas surrounding the spread ratings might apply.

Suppose that we are an e-tailer selling products on the World Wide Web, and suppose that we collect user preference scores in a manner similar to Amazon.com—i.e., each product is scored by a user on a five-star scale.

★	★	★	★	★	(highest recommendation)
★	★	★	★		
★	★	★			$\cdots$
★	★				
★					(lowest recommendation)

The goal is to turn these user scores into a rating system that we can use to suggest highly rated products similar to the ones that the shopper is currently viewing or buying. It is not possible to effectively leaf through a large inventory catalog with a Web browser, especially if one is not sure of what they want or what's in the catalog, so a large portion of the success of online retailing depends on the degree to which a company can be effective in recommending products that customers can trust. It's simple,

Build a better product rating system $\Longrightarrow$ Generate greater trust $\Longrightarrow$ Make more sales!

Building a simple product rating system from user recommendations is easily accomplished just by reinterpreting the skew-symmetric score-differential matrix $\mathbf{K}$ on page 118. For example, suppose that

$$\mathcal{P} = \{p_1, p_2, \ldots, p_n\},$$

is the set of products in our inventory, and let $\mathcal{U}_i$ and $\mathcal{U}_j$ respectively denote the set of all users that have evaluated products p_i and p_j, so $\mathcal{U}_i \cap \mathcal{U}_j$ is the set of all users that have evaluated both of them. When comparing p_i and p_j ($i \neq j$), let $n_{ij} = \#(\mathcal{U}_i \cap \mathcal{U}_j)$ be the number of users that evaluate both of them, and define the "score" that p_i makes against

p_j, to be the average star rating

$$S_{ij} = \begin{cases} \dfrac{1}{n_{ij}} \displaystyle\sum_{h \in \mathcal{U}_i \cap \mathcal{U}_j} (\# \text{ stars given to } p_i \text{ by user } h) & \text{if } n_{ij} \neq 0, \\ 0 & \text{if } n_{ij} = 0. \end{cases} \tag{10.1}$$

Define, the "score difference" between products p_i and p_j to be

$$\begin{aligned} k_{ij} &= S_{ij} - S_{ji} \\ &= \frac{1}{n_{ij}} \sum_{h \in \mathcal{U}_i \cap \mathcal{U}_j} [(\# \text{ stars given to } p_i \text{ by } h) - (\# \text{ stars given to } p_j \text{ by } h)], \quad (10.2) \end{aligned}$$

and, just as in (9.4), let

$$\mathbf{K} = [k_{ij}] = [S_{ij} - S_{ji}] \tag{10.3}$$

be the skew-symmetric matrix of score differences. The development on page 120 produces the following conclusion regarding the best ratings that can be produced by averaging a given set of "star scores."

The Best Star Rated Products

For a given set of star scores, the best (in the sense described on page 118) set of ratings that can be derived by averaging star scores is given by the centroid

$$\mathbf{r} = \mathbf{K}\mathbf{e}/n,$$

where $\mathbf{K}$ is the skew-symmetric matrix of average star-score differences given in (10.3).

For example, suppose ten users $\mathcal{U} = \{h_1, h_2, \ldots, h_{10}\}$ have evaluated four products $\mathcal{P} = \{p_1, p_2, p_3, p_4\}$ using a five-star scale, and suppose that their individual evaluations (number of stars) are accumulated in the matrix

	p_1	p_2	p_3	p_4
h_1	4	2	2	
h_2	3	1		2
h_3	1		2	
h_4	2	4	2	
h_5		3	2	5
$\star\text{-matrix} = \quad h_6$	2	3		
h_7		4	1	3
h_8	3	1	1	
h_9		3	2	5
h_{10}	2		2	4

Notice that there are several missing entries in this matrix because not all users evaluate all products. In fact, the desire to extrapolate values for missing entries was the motivation behind the famous Netflix contest [56]. When the scores S_{ij} defined by (10.1) are arranged in a matrix, the resulting score matrix is

$$\mathbf{S} = \begin{pmatrix} 0 & 14/5 & 12/5 & 5/2 \\ 11/5 & 0 & 17/6 & 11/4 \\ 9/5 & 10/6 & 0 & 7/4 \\ 6/2 & 15/4 & 17/4 & 0 \end{pmatrix}.$$

The skew-symmetric matrix $\mathbf{K}$ of score differences defined by (10.3) is

$$\mathbf{K} = \begin{pmatrix} 0 & 3/5 & 3/5 & -1/2 \\ -3/5 & 0 & 7/6 & -1 \\ -3/5 & -7/6 & 0 & -10/4 \\ 1/2 & 1 & 10/4 & 0 \end{pmatrix}.$$

Therefore, the star ratings derived from user evaluations is

$$\mathbf{r} = \frac{\mathbf{K}\mathbf{e}}{4} = \begin{pmatrix} 7/40 \\ -13/120 \\ -16/15 \\ 1 \end{pmatrix} \approx \begin{pmatrix} .175 \\ -.1083 \\ -1.066 \\ 1 \end{pmatrix},$$

so the products rank (from highest to lowest) in the order $\{p_4, p_1, p_2, p_3\}$.

Direct Comparisons

Using the average star scores (10.2) to define the "score difference" between two products is a natural approach to adopt, but there are countless other ways to define score differences. For example, a retailer might be more interested in making direct product comparisons by setting S_{ij} to be the proportion of people in $\mathcal{U}_i \cap \mathcal{U}_j$ who prefer p_i over p_j. Such comparisons could be determined from user star scores if they were available, but direct comparisons are often obtained from old-fashion market surveys or point-of-sales data. To be precise, let

$$\delta_{ij}(h) = \begin{cases} 1 & \text{when user } h \text{ prefers } p_i \text{ over } p_j, \\ 0 & \text{when user } h \text{ does not prefers } p_i \text{ over } p_j, \end{cases} \tag{10.4}$$

and define

$$S_{ij} = \begin{cases} \dfrac{1}{n_{ij}} \displaystyle\sum_{h \in \mathcal{U}_i \cap \mathcal{U}_j} \delta_{ij}(h) & \text{if } n_{ij} \neq 0, \\ 0 & \text{if } n_{ij} = 0. \end{cases} \tag{10.5}$$

If $\mathbf{K}$ is reinterpreted to be the skew-symmetric matrix of direct-comparison score differences

$$\mathbf{K} = [k_{ij}] = [S_{ij} - S_{ji}], \tag{10.6}$$

then the results in (9.4) on page 120 produce the following statement concerning the best "direct comparison" product ratings.

The Best Direct-Comparison Ratings

The best (in the sense described on page 118) product ratings that can be derived from direct comparisons is the centroid

$$\mathbf{r} = \mathbf{K}\mathbf{e}/n, \tag{10.7}$$

where $\mathbf{K}$ is the skew-symmetric matrix of direct-comparison score differences given by (10.6).

For example, suppose ten users $\mathcal{U} = \{h_1, h_2, \ldots, h_{10}\}$ are involved in making direct comparisons of four products $\mathcal{P} = \{p_1, p_2, p_3, p_4\}$, and suppose that the binary comparison results described in (10.4) are accumulated in the matrix

$$
\delta = \begin{array}{c} \\ h_1 \\ h_2 \\ h_3 \\ h_4 \\ h_5 \\ h_6 \\ h_7 \\ h_8 \\ h_9 \\ h_{10} \end{array}
\begin{array}{cccc} p_1 & p_2 & p_3 & p_4 \\ \end{array}
\left(
\begin{array}{cccc}
1 & & 0 & \\
1 & & & 0 \\
0 & & 1 & \\
0 & 1 & & \\
& 0 & & 1 \\
0 & 1 & & \\
& 1 & 0 & \\
1 & & 0 & \\
& 0 & & 1 \\
0 & & & 1
\end{array}
\right). \tag{10.8}
$$

This means, for instance, that user h_1 compares products p_1 with p_3 and selects p_1, while user h_5 compares p_2 with p_4 and chooses p_4, etc. (The users need not be distinct—e.g., user h_1 could be the same as user h_5.) When the scores S_{ij} defined by (10.5) are arranged in a matrix, the resulting score matrix is

$$
\mathbf{S} = \begin{pmatrix}
0 & 0 & 2/3 & 1/2 \\
1 & 0 & 0 & 0 \\
1/3 & 0 & 0 & 0 \\
1/2 & 1 & 0 & 0
\end{pmatrix}.
$$

Consequently, the skew-symmetric matrix $\mathbf{K}$ of direct-comparison score differences is

$$
\mathbf{K} = \begin{pmatrix}
0 & -1 & 1/3 & 0 \\
1 & 0 & 0 & -1 \\
-1/3 & 0 & 0 & 0 \\
0 & 1 & 0 & 0
\end{pmatrix},
$$

and thus the direct-comparison product ratings derived from (10.7) is

$$
\mathbf{r} = \frac{\mathbf{Ke}}{4} = \begin{pmatrix}
-2/12 \\
0 \\
-1/12 \\
3/12
\end{pmatrix}. \tag{10.9}
$$

In other words, the products rank (from highest to lowest) in the order $\{p_4, p_2, p_3, p_1\}$.

Direct Comparisons, Preference Graphs, and Markov Chains

Another way to derive ratings based on direct comparisons between products is to build a *preference graph*. This is a directed graph with weighted edges in which the nodes of the graph represent the products in $\mathcal{P} = \{p_1, p_2, \ldots, p_n\}$, and a weighted edge from p_i to p_j represents, in some sense, the probability of favoring product p_j given that a person is currently "using" product p_i. While there are countless many ways to design user surveys to determine these probabilities, they can be constructed on the basis of direct comparisons such as those in the previous example (10.8).

1. For each product p_i, list all users $\mathcal{H}_i = \{h_{i_1}, h_{i_2}, \ldots, h_{i_k},\}$ who evaluated p_i.

2. If there is a user in $\mathcal{H}_i$ that prefers product p_j, then draw an edge from p_i to p_j.

3. The weight (or probability) q_{ij} associated with the edge from p_i to p_j is the proportion of users in $\mathcal{H}_i$ that prefer product p_j. In other words, if n_{ij} is the number of users in $\mathcal{H}_i$ that prefer product p_j. then

$$q_{ij} = \frac{n_{ij}}{\#\mathcal{H}_i} = \frac{n_{ij}}{k}.$$

This preference graph defines a Markov chain [54, page 687], and rating the products in $\mathcal{P}$ can be accomplished by analyzing a random walk on this graph to determine the proportion of time spent at each node (or product). This is in essence the same procedure used in Chapter 6 on page 67. The idea is to mathematically watch a "random shopper" move forever from one product to another in the preference graph—if the shopper is currently at product p_i, then the shopper moves to product p_j with probability q_{ij}. The rating value r_i for product p_i is the proportion of time that the random shopper spends at that product.

If there is sufficient connectivity in the preference graph, then the proportion of time that a random shopper spends at product p_i (i.e., the rating value r_i) is the ith component of the vector $\mathbf{r}$ that satisfies the equation.

$$\mathbf{r}^T\mathbf{Q} = \mathbf{r}^T, \quad \mathbf{r}^T\mathbf{e} = 1, \quad \text{where} \quad \mathbf{Q} = \begin{pmatrix} q_{11} & q_{12} & \cdots & q_{1n} \\ q_{21} & q_{22} & \cdots & q_{2n} \\ \vdots & \vdots & \ddots & \vdots \\ q_{n1} & q_{n2} & \cdots & q_{nn} \end{pmatrix}.$$

This vector $\mathbf{r}$ that defines our ratings is also called the *stationary probability vector* for the Markov chain (or random walk). This approach to rating and ranking is in fact the foundation of Google's PageRank—see [49].

For example, consider the direct comparisons given in the matrix δ in (10.8) that are produced when ten users $\{h_1, h_2, \ldots, h_{10}\}$ evaluated four products $\{p_1, p_2, p_3, p_4\}$. From the data in δ we see that

$$p_1 \text{ is evaluated by users } \{h_1, h_2, h_3, h_4, h_6, h_8, h_{10}\} = \mathcal{H}_1,$$
$$p_2 \text{ is evaluated by users } \{h_4, h_5, h_6, h_7, h_9\} = \mathcal{H}_2,$$
$$p_3 \text{ is evaluated by users } \{h_1, h_3, h_8\} = \mathcal{H}_3,$$
$$p_4 \text{ is evaluated by users } \{h_2, h_5, h_7, h_9, h_{10}\} = \mathcal{H}_4.$$

Furthermore, we can also see from δ that of the seven users who evaluated p_1, three of them preferred p_1; two of them preferred p_2; two of them preferred p_3; and one of them preferred p_4. Consequently, there are four edges (paths) leaving node p_1 in the preference graph shown in Figure 10.1 with respective weights $q_{11} = 3/7$; $q_{12} = 2/7$; $q_{13} = 2/7$; and $q_{14} = 1/7$, and thus the first row of $\mathbf{Q}$, shown below in (10.10), is determined. Similarly, the edges (paths) leaving the second node p_2 and the entries in the second row of $\mathbf{Q}$ are derived by observing that of the five users who evaluated p_2, none preferred p_1; three preferred p_2; none preferred p_3; and two preferred p_4, so $q_{21} = 0$; $q_{22} = 3/5$; $q_{23} = 0$; and $q_{24} = 2/5$.

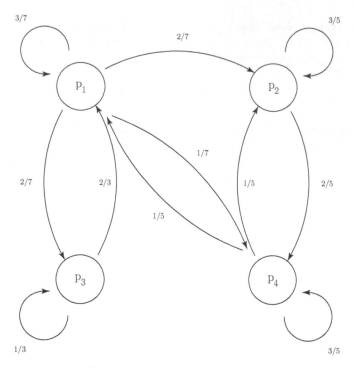

Figure 10.1 Product preference graph

Use the same reasoning to complete the preference graph and to generate the third and fourth rows of

$$\mathbf{Q} = \begin{pmatrix} 3/7 & 2/7 & 2/7 & 1/7 \\ 0 & 3/5 & 0 & 2/5 \\ 2/3 & 0 & 1/3 & 0 \\ 1/5 & 1/5 & 0 & 3/5 \end{pmatrix}. \tag{10.10}$$

The Markov rating vector $\mathbf{r}$ is obtained by solving the five linear equations defined by

$$\mathbf{r}^T(\mathbf{I} - \mathbf{Q}) = \mathbf{0} \quad \text{and} \quad \mathbf{r}^T\mathbf{e} = 1$$

for the four unknowns $\{r_1, r_2, r_3, r_4\}$. The result is

$$\mathbf{r} = \begin{pmatrix} 7/30 \\ 1/3 \\ 1/10 \\ 1/3 \end{pmatrix} \approx \begin{pmatrix} .2333 \\ .3333 \\ .1 \\ .3333 \end{pmatrix}, \tag{10.11}$$

so our four products are ranked (from highest to lowest) as $\{\{p_4, p_2\}, p_1, p_3\}$, where p_4 and p_2 tied.

Centroids vs. Markov Chains

There are several problems with the Markov chain (or random walk) approach to rating. First, when there are a lot of states (or products in this case), there is often insufficient connectivity in the graph to ensure that the solution of $\mathbf{r}^T\mathbf{Q} = \mathbf{r}^T$, $\mathbf{r}^T\mathbf{e} = 1$ (i.e., the ratings vector) is well defined. For such cases artificial information or perturbations must be somehow forced into the graph to overcome this problem, and this artificial information necessarily pollutes the purity of the "real" information—see [49]. Consequently, the resulting ratings and rankings are not true reflections of the actual data. However, this is a compromise that must be swallowed if Markov chains are to be used. On the other hand, using a centroid method requires no special structure in the data, so no artificial information need be introduced to produce ratings. Furthermore, even if the graph can be somehow perturbed or altered to guarantee that a well defined stationary probability vector exists, computing r can be a significant task in comparison to computing centroids, which is relatively trivial.

Conclusion

The centroid rankings (10.9) on page 130 and the Markov ratings (10.11) are both derived from the same direct comparisons data in (10.8). The centroid method ranks the products in the order $\{p_4, p_2, p_3, p_1\}$, while the Markov method yields the ranking as $\{\{p_4, p_2\}, p_1, p_3\}$. It is dangerous to draw conclusions from small artificial examples like these, but you can nevertheless see that the centroid rankings are in the same ball park as the Markov rankings. What is clear from this example is that the centroid rankings are significantly easier to determine, so if a centroid method can provide you with decent information, then they should be the method of choice.

By The Numbers —

$24 =$ minutes remaining when team BelKor edged out team Ensemble
          to win the $\$1,000,000$ Netflix prize for improving user preference
          predictions by 10%.

                    — July 26, 2009 at 18:18:28 UTC.

                                        — netflixprize.com

Chapter Eleven

Handling Ties

Companies such as Amazon, Netflix, and eBay solicit and collect data on user behavior, which results in enormous databases with records of user ratings of products and services. One common goal of analyzing this data is to create rankings of items, which may then be used as part of a recommendation system. Typically, the first step in creating such a ranking is to transform the user ratings into pair-wise comparisons. There are several methods for doing this. See, for instance, [33] and the material in Chapter 10. Every transformation begins by creating head-to-head contests between pairs of items. For example, in a database of user ratings about movies, the (i, j)-element of a possible pair-wise matrix could contain the number of times movie i beat movie j in their contests, where a contest occurs each time a user has rated both movies. As a result, it is clear that ties are prevalent in many database applications.

This chapter discusses ranking in the presence of ties in the input data, and many of the results are compilations of those in [20]. Ties can also exist in the output (the final rankings)—more on this topic appears in Chapter 15. The issue of how to handle ties in the input data has received little attention despite the fact that this decision can have a significant impact on ranking results. Many ranking methods handle ties by simply ignoring them. Here are three compelling reasons for a more careful consideration of ties.

- Depending on the application, ties can comprise a sizeable portion of the data. For most NHL seasons, 9–13% of games end in ties. Ties are much more prevalent in database applications. For example, when ranking movies rated by users, it is typical to find that up to 50% of the "game" data in this context are ties. In this scenario, a game occurs when two movies have both been rated by the same user. The movie receiving the higher rating wins that "game."

- Some methods, such as the Massey method, incorporate ties while others, such as the Colley method, ignore them. Thus, when comparing rankings produced by various methods, it is important to make the comparison fair by treating ties uniformly across all methods. We advocate accounting for, rather than ignoring, ties. When ties are accounted for, a low-ranked team that ties a high-ranked team is rewarded for its effort while the high-ranked team is penalized. This reaction to the tie has more intuitive appeal than treating it as a non-event.

- There are some sports where ties are rare or impossible due to overtime or shoot-out rules. In such sports, it sometimes happens that two teams were very evenly matched for the large duration of the game, yet one emerged the victor due to a last minute heroic. On page 142, we induce ties for these near-tied events and other near-tied events such as those with very small point differentials. We show that these induced ties can improve the predictive power of the ranking results. In applications such as the BCS rankings of college football teams, small changes in the ranking results can have huge impacts on universities, given the revenue associated with bowl invitations.

Input Ties vs. Output Ties

In general, ties have received very little attention in the ranking literature. Here we distinguish between two types of ties. *Input ties* are ties in the input data (i.e., games that end in a tied score) and *output ties* that are ties in the output data (i.e., when a ranking method outputs a ranking that contains ties). Some work has been done regarding output ties, particularly in the context of rank aggregation [3, 4, 23, 27, 50]. Rank aggregation is related to, yet different from, ranking. Rank aggregation, which is also known as consensus ranking, creates one ranking that is an aggregate of several other rankings. See Chapters 14 and 15 for more on rank aggregation. In contrast, in this chapter, we are interested in input ties and their effect on ranking, not rank-aggregation, methods.

Incorporating Ties

For a ranking application in which the data contains ties, one must be very careful when comparing rankings produced by different methods. Some methods, such as the Massey and OD methods, very naturally account for ties, while others, such as the Colley and some Markov variants, ignore them. All methods either already incorporate or can be modified to use information from ties. Consequently, in order to make fair comparisons across methods, for each method, one must derive two models: one when ties are used and another when ties are omitted.

The Colley Method

Recall from Chapter 3 that the Colley method can be succinctly summarized with one linear system, $\mathbf{Cr} = \mathbf{b}$, where $\mathbf{r}_{n \times 1}$ is the unknown Colley rating vector, $\mathbf{b}_{n \times 1}$ is the right-hand side vector defined as $b_i = 1 + \frac{1}{2}(w_i - l_i)$ and $\mathbf{C}_{n \times n}$ is the Colley coefficient matrix defined as

$$\mathbf{C}_{ij} = \begin{cases} 2 + t_i & i = j, \\ -n_{ij} & i \neq j. \end{cases}$$

The scalar n_{ij} is the number of times teams i and j played each other, t_i is the total number of games played by team i, w_i is the number of wins for team i, and l_i is the number of losses for team i.

Since the Colley method only accounts for wins and losses, ties are, in effect, non-events that are ignored. Thus, we now refer to the above model as the *Colley-no-ties method*. In order to incorporate ties, we must modify the above model, creating the *Colley-ties method*.

Incorporating ties into the Colley method requires some thought as to the meaning of a tied outcome. One interpretation of a tie argues that, if the two teams were to meet again, there would be an equal chance that either team would win. Or complementarily, we could say that there is an equal chance that either team would lose. Thus, to include ties we count a tie between teams i and j as two games, dividing the usual weight of 1 equally between the two artificial games. In one artificial game, team i beats team j with a weight of 1/2 while in the other team j beats team i. This interpretation of a tie leaves the $\mathbf{b}$ vector unaltered and modifies the $\mathbf{C}$ matrix in an obvious way (i.e., for a tie between teams i and j, c_{ij} and c_{ji} decrement by 1 and c_{ii} and c_{jj} increment by 1). One nice consequence of this interpretation of ties is that the Colley conservation property, which states that the sum of all Colley ratings is $n/2$, is maintained.

The Massey Method

Recall from Chapter 2 that the Massey method can be succinctly summarized with the linear system

$$\mathbf{Mr} = \mathbf{p},$$

where the Massey coefficient matrix $\mathbf{M}$ is related to the Colley matrix $\mathbf{C}$ by the formula $\mathbf{M} = \mathbf{C} - 2\,\mathbf{I}$ so that

$$\mathbf{M}_{ij} = \begin{cases} t_i & i = j, \\ -n_{ij} & i \neq j. \end{cases}$$

The Massey right-hand side vector $\mathbf{p}$ is a vector of cumulative point differentials. That is, p_i is the total number of points team i scored on all opponents minus the total number of points opponents scored against team i. Since $rank(\mathbf{M}) = n - 1$, an adjustment is made to ensure nonsingularity. Any row of $\mathbf{M}$ is replaced with a row of all 1s and the corresponding entry in $\mathbf{p}$ is set to 0. This new constraint forces the ratings to sum to 0. Following Massey's advice, we apply this nonsingularity adjustment to the last row, creating an adjusted linear system which we denote $\bar{\mathbf{M}}\mathbf{r} = \bar{\mathbf{p}}$.

The above Massey method naturally incorporates ties. For a tied event between teams i and j, the m_{ij} and m_{ji} entries increment by one, yet the information in the point differential vector $\mathbf{p}$ remains unchanged. Thus, we refer to the above method as the *Massey-ties method*.

In order to properly analyze the methods there needs to be a comparable *Massey-no-ties method* that does not incorporate ties in creating the rating. To accomplish this with minimal change, alter the Massey method to ignore any game that ends in a tie. That is, we do not update the $\mathbf{M}$ matrix or $\mathbf{p}$ vector when a tie occurs between teams i and j. This behaves similar to the Colley-no-ties method in that these ties are considered non-events.

The Markov Method

The key to the Markov method of Chapter 6 is voting. A weaker team casts votes for each stronger team it faces, and because there are several methods for vote casting, the Markov method has several variants. In this chapter, we consider the following three.

1. A losing team casts one vote for each team it is beaten by.

2. A losing team casts a number of votes equal to the number of points it was beaten by.

3. For each game, both winning and losing teams cast a number of votes equal to the number of points given up.

Regardless of the vote casting method, the result is a team-team matrix that is row-normalized in order to enforce stochasticity, which creates the transition matrix for a Markov chain. The stationary vector of this chain is the Markov rating vector. As discussed in Chapter 6, the elements of the stationary vector correspond to the proportion of time one would visit a particular team if one took a random walk on the graph defined by the voting matrix. The higher the rating (i.e., the higher the value in the stationary vector), the more frequent the visits to that team, and hence, the greater the relative importance of that team.

It will soon become apparent that the method of vote casting is the focal point in our discussion of ties within the Markov method. Thus, we pursue each vote casting method in turn.

1. *(Loser votes with loss.)* As originally designed, this Markov method ignores ties. However, accommodating ties is easy. In fact, we borrow from the Colley-ties method—for each tied outcome, we let both teams vote with half a vote.

2. *(Loser votes with point differential.)* As originally designed, this Markov method also does not count ties. Unfortunately, unlike the above voting Markov method, ties are not easily accommodated. In fact, if the two teams in the tied event have not played each other more than once, a tied outcome cannot be recorded. If the two tying teams have played more than once, then ties can be handled. Suppose teams i and j played each other twice. Team i beat team j by 4 points in the first game while the second game ended in a tie. Information from multiple games between two teams is typically recorded in the Markov matrix as the *average*, rather than cumulative, point differential. Consequently, in this example, the (i, j)-element would be 2 with a *Markov-ties method,* as opposed to 4 in a *Markov-no-ties method*. However, since the number of games between pairs of teams is not constant, we advise against using this method of vote casting. It makes for an unfair comparison in a dataset that contains ties.

3. *(Both winner and loser vote with points.)* As originally designed, this Markov method naturally accommodates ties. In fact, to ignore ties, one would have to remove any tied event from the dataset.

The OD, Keener, and Elo Methods

Some ranking methods such as the three of this section naturally account for ties. The OD method of Chapter 7 computes two ratings for each team, an offensive and a defensive value, that can then be aggregated into one rating value for each team. As described in (7.7) and (7.8), the two OD rating vectors $\mathbf{o}$ and $\mathbf{d}$ are computed by the alternating refinement process

$$\mathbf{o}_k = \mathbf{A}^T \mathbf{d}_{k-1}^{\div},$$

and

$$\mathbf{d}_k = \mathbf{A} \mathbf{o}_k^{\div}, \qquad \text{for } k = 1, 2, 3, \dots,$$

where $\mathbf{A} = [a_{ij}]$ is a "score matrix" in which a_{ij} is the score that team j generated against team i (average the results if the teams met more than once, and set $a_{ij} = 0$ if the two teams did not play each other). After convergence, $\mathbf{o}$ and $\mathbf{d}$ are aggregated to create an overall rating vector $\mathbf{r}$. While there are several aggregation choices, the simple rule $\mathbf{r} = \mathbf{o}/\mathbf{d}$ works well in practice.

Because the OD method uses point score data, it naturally accommodates ties. Thus, we call the above method the *OD-ties method*. In order to create an *OD-no-ties method*, set the point scores of tied games to 0.

In a similar fashion, most variants of the Keener method of Chapter 4 also naturally account for ties since they too are built from point score data. Ties are built into the Elo rating system of Chapter 5, which is important given the frequency of tied outcomes in soccer and chess, two major applications of the Elo method.

Theoretical Results from Perturbation Analysis

In this section, we use tools from perturbation analysis to quantify the effect that ties can have on Colley rankings. While similar results can be derived for other methods, to avoid redundancy we use just one method, the Colley method, to make our main points.

We begin with a season of data that up to this point has no ties. Next we compare the rankings produced by the Colley-no-ties method and Colley-ties method when one additional game, *which results in a tie*, is played. Before this perturbation, both methods produce the same ranking $\mathbf{r} = \mathbf{C}^{-1}\mathbf{b}$. After this perturbation, the Colley-no-ties method ignores this additional game, and thus, the ranking is unchanged. On the other hand, the ranking of the Colley-ties method will change as a result of this perturbation. Fortunately, we can precisely examine this change with perturbation analysis. In particular, the tied event manifests as a rank-one update to the Colley coefficient matrix while the right-hand side vector $\mathbf{b}$ remains unchanged.

Let $\tilde{\mathbf{C}}$ represent the Colley matrix for the Colley-ties method after the perturbation. This matrix $\tilde{\mathbf{C}}$ is related to the pre-perturbation matrix $\mathbf{C}$ by the formula

$$\tilde{\mathbf{C}} = \mathbf{C} + (\mathbf{e}_i - \mathbf{e}_j)(\mathbf{e}_i - \mathbf{e}_j)^T.$$

Using the Sherman-Morrison rank-one update formula and some M-matrix properties [54] of the Colley matrix, we can express $\tilde{\mathbf{r}}$ in terms of $\mathbf{C}$. Specifically,

$$\begin{aligned}
\tilde{\mathbf{r}} &= \tilde{\mathbf{C}}^{-1}\mathbf{b} \qquad\qquad\qquad\qquad\qquad\qquad\qquad\qquad\qquad (11.1)\\
&= (\mathbf{C} + (\mathbf{e}_i - \mathbf{e}_j)(\mathbf{e}_i - \mathbf{e}_j)^T)^{-1}\mathbf{b}\\
&= \mathbf{r} \quad (\frac{r_i - r_j}{1 + [C^{-1}]_{ii} - 2[C^{-1}]_{ij} + [C^{-1}]_{jj}})\mathbf{C}^{-1}(\mathbf{e}_i - \mathbf{e}_j).
\end{aligned}$$

We draw two conclusions from the above equation.

- A single tied game can affect the ratings of every team. This is because the ith and j^{th} columns of $\mathbf{C}^{-1}$ are almost always completely dense.

- Suppose, before the perturbation, that team i was ranked above team j so that $r_i - r_j > 0$. Then with the additional event of a tie between teams i and j, the rating of team i, $\tilde{r}_i$, drops while the rating of team j, $\tilde{r}_j$, rises. This is a consequence of the diagonal dominance and nonnegativity of $\mathbf{C}^{-1}$.

Next, we wonder if the post-perturbation drop in rating for team i is enough to cause a drop in ranking as well. Mathematically, if $\tilde{r}_i < \tilde{r}_{i+1}$, then team i will drop one rank position as a result of the tie. In order to quantify the conditions under which team i's post-perturbation ranking drops, we define $\epsilon = r_i - r_{i+1}$ as the difference in the pre-perturbation ratings of teams i and $i+1$. Further dissecting Equation (11.1), we can conclude that team i drops one rank position, i.e.,

$$\tilde{r}_i < \tilde{r}_{i+1} \quad \Longleftrightarrow \quad \epsilon < \frac{(r_i - r_j)([C^{-1}]_{ii} - [C^{-1}]_{ij} - [C^{-1}]_{i+1,i} + [C^{-1}]_{i+1,j})}{1 + [C^{-1}]_{ii} - 2[C^{-1}]_{ij} + [C^{-1}]_{jj}}.$$

From this statement, we make two more observations.

- If teams i and j are far apart in rating (i.e., $r_i \gg r_j$), then team i is likely to drop in rank.

- On the other hand, the closer teams i and j are in rating, the less likely it is for teams i and j to change in rank as a result of the tied event.

In this section, we have quantified the impact of a lone tied event on the two Colley methods, the Colley-no-ties and Colley-ties methods. In summary, we have found that *the presence of a single tied event can change both the rating and the ranking vectors*. Real applications often contain many more ties than a lone tied event, which further amplifies our theoretical finding that the handling of ties can have an effect on the ratings and rankings produced by the Colley method. Perturbation analysis for the other ranking methods reaches similar conclusions. In a sense, the theory of this section is a sanity check for the experiments of the next section, which echo our theoretical findings.

Results from Real Datasets

In this section, we move from the theoretical tool of perturbation analysis to real applications. Here we have one simple goal—to show that the handling of ties can have a significant impact on ranking results from real data.

Ranking Movies

Our first application comes from movies. The need for ranking database items such as movies or books or products continues to grow with data storage capabilities. Recommendation systems, such as Netflix or Amazon, use database entries on purchase history and user ratings to make recommendations to users. These recommendations are derived from clustering and ranking algorithms [33]. Thus, because ties are known to be quite prevalent in database applications [27], their proper handling is essential for accurate ranking results.

In this section, we use a well-studied dataset called MovieLens, which contains 100,000 integer ratings (1 through 5) of 943 users regarding 1,682 movies. This data creates a $1,682 \times 1,682$ matrix, that is sparse since most users rate only a small proportion of the movies. In order to rank movies, we create artificial matchups between each pair of movies. A "game" occurs between movie i and movie j every time a user rated both movies and the user ratings become the "scores" for the two movies. With these definitions, we find that the MovieLens dataset contains 32% ties in the user ratings. To produce the results shown in the following diagram,

Colley rankings for 20 movies: ties vs. no ties

we ranked all 1682 movies in the MovieLens data set and selected 20 popular titles to display. The left side represents the rankings having incorporated ties and the right side shows the rankings having omitted any ties. Since ties constitute about a third of the data, the two rankings, as expected, are different.

Proponents of the "ignore ties" strategy argue that ties are not worth modeling since either they are rare or likely affect only lower ranked items. However, the above diagram shows that ties can affect items at any rank position, including the top few positions. Clearly, the modeling of ties can have a dramatic impact on ranking results.

Ranking NHL Hockey Teams

We turn to sports for our second application. Here we study NHL hockey, a low-scoring sport in which ties are a regular occurrence. Ties typically comprise anywhere from 9–13%

of the outcomes in a given season. Like the MovieLens data, we find that accounting for ties as opposed to ignoring them produces two disparate rankings.

The natural question is: do methods that account for ties produce better rankings? It depends on how you define better. Often better is related to predictive capability—i.e., *foresight,* where the question is: do methods with ties outperform methods without ties when it comes to predicting the outcomes of future games? Better can also be related to retrodictive capability—i.e., *hindsight,* where the question is: do methods with ties outperform methods without ties when it comes to matching the outcomes of past games? The following graph shows the foresight performance of the Colley, Massey, and Markov methods considering ties vs. no ties for NHL seasons 2001–2005.

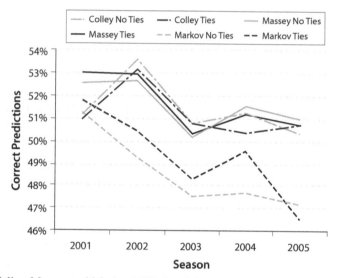

Colley, Massey, and Markov NHL foresight performance: ties vs. no ties

This graph does not point to a winner in the ties vs. no-ties debate.[1] We tested several other seasons of other sports, including basketball and football, and came to the following conclusion. *While incorporating ties into a ranking method sometimes slightly improves the predictive performance, it rarely hurts.* This rather lukewarm endorsement of ties prompted us to study the effect of induced ties, which are the subject of the next section and provide our most interesting finding in the ties vs. no-ties debate.

Induced Ties

While some sports like hockey or soccer admit ties, other sports use overtime or sudden death rules to break ties that occur at the end of regulation time. In addition, there are some non-overtime events that appear to be won almost arbitrarily. The two teams are gridlocked in a tie or near tie for almost the entire game, then some last minute heroic breaks the tie.

---

[1]We dug deeper by examining the locations at which the two methods differed in their prediction and tried to determine if one method did a better job predicting at top-ranked positions, yet failed at lower-rank positions. However, this too drew no clear conclusion in the ties vs. no ties debate.

Basketball even has a phrase for such heroics—a "buzzer beater" refers to a shot that wins the game in the waning seconds. In such situations, it can be argued that a tie would be a more appropriate measure of the two teams' strengths. Consequently, in this section, we *induce* ties in these instances and study the effect of our changes. For example, in NFL football we induce a tie for every game that ended with a final point differential of three or fewer points. The following three graphs show the effect of induced ties on the ten NFL seasons from 1999 to 2008 for the three ranking methods of Colley, Massey, and Markov.

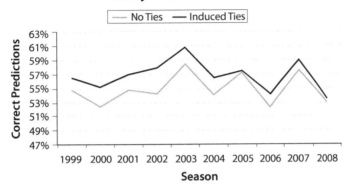

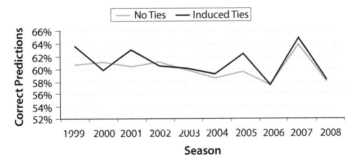

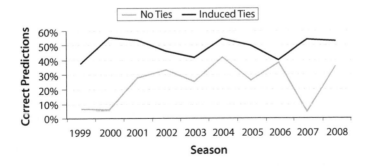

The previous three graphs show that the induction of ties clearly improves the predictive performance of these three ranking methods. The results were similar for the other ranking methods and other datasets such as college football.

We also studied the effect of induced ties in college basketball. While football has a rather obvious margin of victory for which to induce a tie (the 3 points scored by a field goal), basketball has a less obvious margin, so we varied the number of points required to induce a tie from 0 to 6. The results for the Colley method are shown below.

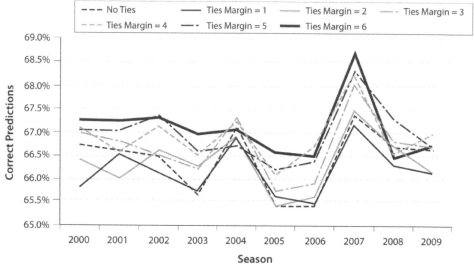

The no-ties results are the dashed line while all induced-ties methods are represented by solid lines. The solid lines vary in darkness from the light gray of the induced-ties method for the smallest margin of victory (one point) to the black of the largest margin of victory (6 points). Most margins of victory for the induced-ties methods outperform their no-ties versions. While it is hard to determine which margin of victory gives the best improvement in terms of the predictive performance, we do see a trend that the larger margins of victory generally outperform the lower margins of victory. We save a more precise study of the optimal margin of victory for inducing ties for future work.

In summary, the experiments of this section show that there is a clear winner in the ties vs. no-ties debate. The induced-ties method regularly outperforms the no-ties method as a predictive scoring measure.

Summary

Given the availability of large datasets in the modern world of computing and the frequent need to rank data, the fact that one may wish to use sports ranking methods in which input data may have ties, and the potential impact on incorporating ties into such results, one should carefully consider using methods that include ties. This chapter discussed how to include ties in several common ranking methods. Further, we introduced the idea of induced ties, which can improve the predictive performance of rankings.

Some questions remain. For example, what is the most effective way to induce ties for specific sports? Perhaps it makes sense to induce a tie in a low-scoring sport like soccer when the winning goal was scored in the last few minutes of regulation play. Of course, this raises the question of how many minutes should be viewed as "the last few" of a game. In basketball, a game can be close until the losing team begins fouling, hoping missed shots will reduce the margin of victory with little reduction in the time left in the game. You might ask if inducing ties in this context was beneficial and, presumably, how to automate such a decision among all games since making such decisions game by game is impractical. Other statistics could be also be analyzed—e.g., points per possession rather than score might be integrated in a ranking method and decisions regarding ties.

In addition, are there datasets where one expects ties to be less important? Are there datasets where one expects ties to affect certain locations in the ranking? For what other applications and methods are ties useful?

By The Numbers —

   6 = most number of ties by an NFL football team in one season.
         — Chicago Bears, 1932.

   24 = most number of ties by an NHL hockey team in one season.
         — Philadelphia Flyers, 1969–70.

   2 = fewest number of ties by an NHL team in one season.
         — Boston Bruins, 1929–30.

      — en.wikipedia.org & couchpotatohockey.com

Chapter Twelve

Incorporating Weights

"A preoccupation with the future not only prevents us from seeing the present as it is but often prompts us to rearrange the past." —Eric Hoffer (1902–1983)

Rearranging the past is what this chapter is about. In fact, this is the first chapter that is more art than science. The art of ranking includes the ability to customize a method based on expert or application-specific information. Here's a scenario: Team A lost two early-season games, then went undefeated for the remainder of the season, while team B was undefeated save for two late-season games. In your opinion, which team should be ranked higher? Most people say team A, since "preseason doesn't matter." That intuition sparked the mathematics of this chapter.

In all of the models presented thus far (with the possible exception of the Elo method), every matchup has been weighted equally. While at first blush, it sounds fair that no matchup carries more weight than another, this may not make sound modeling sense. For instance, in sports, shouldn't end-of-season tournament games carry more weight than pre-season or early season games? Shouldn't a team on a hot streak be more likely to win than a team on a cold streak? For webpages, perhaps pages that are updated more often should be more important. As a result, it's natural to try to incorporate such time considerations into the models.

Weighting Ideas

While we emphasize *weighting by time* in this chapter, all sorts of weightings are possible. For example, one could *weight wins at away locations* more heavily than wins at home. Or one could *weight wins against known rivals* more heavily. That's one feature of weightings—they are flexible enough to accommodate any weighting rationale you can dream up.

Four Basic Weighting Schemes

Weighting games is easy in most models. Though there are numerous possibilities and variations on the following, here we present four basic weighting schemes: linear, logarithmic, exponential, and step function weighting. To relay how easy and natural weighting is, we demonstrate weighting with the Markov model. In this case, the entries v_{ij} of the weighted Markov voting matrix $\bar{\mathbf{V}}$ are computed as

$$\bar{v}_{ij} = w_{ij} \, v_{ij},$$

where w_{ij} is a scalar that accounts for the weight of the game played by teams i and j, and v_{ij} is the number of votes team i cast for team j (from the voting options for the Markov model of Chapter 6). The weight w_{ij} is determined by the weighting function. We propose four basic weighting functions, which are displayed in Figure 12.1 and described in turn below. Again, we talk specifically about time weightings, but any weighting or combination of weightings may be used.

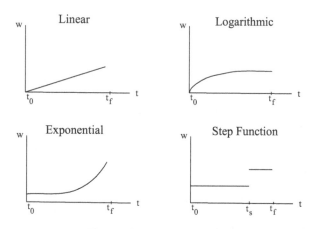

Figure 12.1 Four basic weighting functions

For a linear weighting scheme,

$$w_{ij} = \frac{t - t_0}{t_f - t_0}.$$

The numerator represents the time of the game relative to the season opener. Specifically, it is the number of days since the season opener, which occurred at time t_0. The denominator is the total number of days in the season, which is given by the time of the season finale t_f minus the time of the season opener t_0.

If we want to weight later games in a slowly increasing manner, we use a logarithmic weighting scheme such as

$$w_{ij} = log(\frac{t - t_0}{t_f - t_0} + 1).$$

On the other hand, if we want to dramatically exaggerate the weight of games at the end of the season, we use an exponential weighting scheme

$$w_{ij} = e^{\frac{t - t_0}{t_f - t_0}}.$$

The fourth weighting scheme that we describe in this chapter uses a step function. One possible step function weights games that occur in the last 2 (or 3 or 4 or k) weeks more than other games. One way to add extra weight to games occurring later in the season is to simply double (or triple or quadruple, etc.) count their corresponding votes. This works rather well, particularly for basketball as shown by the experiments of Luke Ingram's M.S. thesis [41]. In this case,

$$w_{ij} = \begin{cases} 1 & \text{if matchup of teams } i \text{ and } j \text{ occurs before } t_s \\ 2 & \text{otherwise,} \end{cases}$$

where t_s is some specific time during the season, such as one month prior to the conference tournament. Of course, a multi-step function could be implemented easily as well. This is precisely what Davidson College undergraduate Erich Kreutzer did for the 2009 March Madness tournament [19]. He created a step function that increased the weight of games every two weeks. His biweekly step function for the Colley method did so well that it got an ESPN score of 1420 (out of a total possible 1600) and was in the 97.3 percentile of all 4.6 million brackets submitted to the annual ESPN Tournament Challenge that year.

Summary of and Notation for the Time Weighting Schemes

w_{ij} weight given to matchup between teams i and j

t_0 time of season opener (e.g., day 1 of season occurs when $t_0 = 0$)

t_f time of final game of season

t_s specific time during season to change step weighting

t time of current game under consideration

Below are four simple methods for weighting a game between teams i and j that occurred at time t.

Linear

$$w_{ij} = \frac{t - t_0}{t_f - t_0}$$

Logarithmic

$$w_{ij} = log(\frac{t - t_0}{t_f - t_0} + 1)$$

Exponential

$$w_{ij} = e^{\frac{t - t_0}{t_f - t_0}}$$

Step

$$w_{ij} = \begin{cases} 1 & \text{if game for } i \text{ and } j \text{ occurs before } t_s \\ 2 & \text{otherwise.} \end{cases}$$

Now let's discuss the details concerning how a weighting scheme can be applied to specific ranking methods.

Weighted Massey

For the Massey model, each game in a season is represented by a row in the original Massey linear system $\mathbf{Xr} = \mathbf{y}$. Each game and thus each row of the Massey system can be weighted according to its temporal occurrence in the season using any chosen weighting scheme. In this case, rather than solving $\mathbf{X}\mathbf{r} = \mathbf{y}$ as a least squares problem, a *weighted least squares* problem is solved. First, a vector $\mathbf{w}$ of weights associated with each game is created. Again, perhaps games are weighted by their time from the beginning of the season so that late season games are more important. Next, $\mathbf{w}$ is transformed into a diagonal matrix $\mathbf{W}$ that has $\mathbf{w}$ along its diagonal. Finally, the weighted normal equations [36]

$$\mathbf{X}^T\mathbf{W}\mathbf{X}\mathbf{r} = \mathbf{X}^T\mathbf{W}\mathbf{y}$$

are solved in order to produce the unique weighted least squares solution to the weighted original system $\mathbf{W}^{1/2}\mathbf{X}\mathbf{r} = \mathbf{W}^{1/2}\mathbf{y}$.

Weighted Colley

To weight the Colley method the off-diagonal elements of the Colley matrix $\mathbf{C}$ are no longer integers for the number of times teams faced each other, instead they are weights associated with the importance of each game. For example, in the unweighted context, if $c_{ij} = 3$, meaning teams i and j faced each other three times, then in the weighted context, c_{ij} is the sum of the weights associated with these three matchups. Similarly, the right-hand side vector $\mathbf{b}$ no longer contains a count of total wins minus total losses, instead it contains weighted counts. In terms of programming, this weighting modification is an easy adjustment to the algorithm.

Weighted Keener

Once a weighting method has been selected, the weighted Keener matrix $\bar{\mathbf{A}}$ is populated according to the formula

$$\bar{a}_{ij} = w_{ij}\, a_{ij},$$

where w_{ij} is the weight of the matchup between teams i and j and a_{ij} is the Keener statistic of choice from Chapter 4 . Once this weighted Keener matrix is available, the remaining steps of the Keener method are executed as usual. Because nonnegativity of the Keener matrix is required, one must use nonnegative weights w_{ij}.

Weighted Elo

The K factor of the Elo method is the method's built-in mechanism for weighting. Recall from Chapter 5 that for soccer the constant K is used to weight wins during World Cup finals ($K = 60$) more than World Cup qualifiers ($K = 40$) more than other tournaments ($K = 30$).

Weighted Markov

Once a weighting method has been selected, the weighted Markov voting matrix $\bar{\mathbf{V}}$ is populated according to the formula

$$\bar{v}_{ij} = w_{ij}\, v_{ij},$$

where w_{ij} is the weight of the matchup between teams i and j and v_{ij} is the number of votes team i cast for team j using one of the voting methods of Chapter 6. Once this weighted voting matrix is available, the remaining steps of the Markov model are executed without modification. That is, the weighted voting matrix $\bar{\mathbf{V}}$ is row normalized and, if necessary, a procedure for handling undefeated teams is applied. At this point, the stochastic Markov matrix $\mathbf{S}$ is available and the stationary vector, i.e., the rating vector $\mathbf{r}$, is computed.

Weighted OD

Weighting the OD method is nearly identical in implementation to the Markov method. That is, weighting is a preprocessing step to the OD method. Specifically, the elements p_{ij} of the OD $\mathbf{P}$ are massaged with the appropriate weighting scheme, then OD is computed as usual using this weighted $\mathbf{P}$ matrix.

Weighted Differential Methods

Time weighting can be incorporated easily into both the rank-differential and the rating-differential methods of Chapter 8. The actual algorithms do not change, only the input data matrix $\mathbf{D}$ changes as it is weighted before normalization according to the techniques discussed on page 147.

ASIDE: Weighting and the March Madness Tournament

Our work with weighting and the annual March Madness basketball tournament began in 2006 and has developed its own history in that time. Each year our weighting work gets more sophisticated as an outgoing group of researchers passes the torch to an incoming group. Here we chronicle the highlights of this work and introduce the main contributors.

- **2006**: For a class project at the College of Charleston, then graduate students Luke Ingram and John McConnell were the first to tackle a March Madness prediction problem. Their goal was to use mathematical techniques to predict games in the tournament. The two discovered ESPN's Tournament Challenge (and their $10,000 prize) and its online automated bracket submission and scoring tool. That year the pair submitted a few variants of the Markov model to ESPN. Because Luke and John believed momentum had strong predictive power, one submission used time-weighting to doubly count games in the month leading up to the tournament. Unfortunately, for Luke and John, whom other classmates had dubbed "The Apostles," 2006 was the Year of the Upset. Many readers may remember the Cinderella run that the 11-seeded George Mason had, making it all the way to the final four. Only a very small crop of fans (mostly G.M. alumni) had George Mason advancing this far. In fact, the fan who did win the ESPN Tournament Challenge that year had mistakenly entered George Mason, meaning instead George Washington. After realizing his mistake, Russell Pleasant submitted his corrected G.W.-favored bracket. Luckily, he did not delete his G.M.-favored bracket from the contest, as that was the one that earned him the $10,000 prize that year.

- **2008**: In the spring of 2008, I [A. L.] assigned the same project to seniors Neil Goodson and Colin Stephenson. These two sports fans and mathematics majors eagerly took up the torch, picking up where Luke and John left off. By the ESPN deadline, our class helped the pair submit over 30 brackets created from rankings of various weightings of the Colley, Massey,

Markov, and OD methods. Neil and Colin studied their brackets well and discovered some newsworthy trends.

Figure 12.2 Neil Goodson and Colin Stephenson

For instance, most models predicted the dramatic upset of #11-seeded Kansas State over #6-seeded USC. Also, as most models predicted Kansas to win the whole tournament over Memphis, Neil and Colin decided to use the Massey method to predict the final score of a game between these two teams. Thus, they predicted Kansas to win in a very close game by just 3 points, a very good prediction given that Kansas won in overtime. The College of Charleston Media folks heard about Neil and Colin's work and checked in regularly for updates as the tournament progressed. Propelled by their fantastic success in early rounds, Neil and Colin found themselves at the center of a media storm. Local media, including several newspaper and television stations, contacted the pair. Then came the national media, including an appearance on CBS's *The Early Show*. Perhaps the most fame-inducing invitation came from Robert Siegel and his *All Things Considered* show on National Public Radio (NPR). Neil had a live phone interview during which he mentioned a few of his predictions, namely the big Kansas State upset and the champion Kansas University. Neil was a natural—striking just the right balance between the technical explanations of the mathematics and the exciting sports results and predictions. Neil's segment received positive responses nationwide. In fact, he was so engaging that several nonsports listeners were so flabbergasted by the eventual accuracy of Neil's prophesies that they felt compelled to call into the show after the tournament had ended. One listener called in to say that the pair did such a great job on their class project that their professor ought to give them an A. Their professor did—not for their spot-on predictions, but for their initiative and work ethic week after week all semester long. For more on the Neil and Colin story including one of their highest-scoring brackets, see the Aside on page 212.

- **2009**: My [A. L.] Operations Research project class is offered only every other year at the College of Charleston. With the 2008 success fresh in mind, I certainly could not let this small fact inhibit our progress and 2009 March Madness fun. Thus, I teamed up with my good colleague, Dr. Tim Chartier of Davidson College, to run a cross-institutional ranking project. Our College of Charleston team consisted of graduate student Kathryn Pedings and undergraduate Ryan Parker. The Davidson team included Dr. Chartier and his undergraduate students Erich Kreutzer and Max Win. The overall team expanded to include Nick Dovidio, a Davidson alumni then at Stanford University, and Dr. Yoshitsugu Yamamoto of Tsukuba University, Japan. Our team truly spanned the country, continent, and globe as Figure 12.3 shows.

Kathryn Pedings   Yoshitsugu Yamamoto

Figure 12.3 Our March Madness team spans the globe

In 2009 we added a few more weightings as well as rankings from several rank-aggregation models (see Chapters 14 and 15). While 2009 was not quite as bracket-friendly as the historic and heralded 2008, we were very pleased with our progress. Several models scored in the 95th and 97th percentiles of all 4.6 million brackets submitted to the ESPN Challenge that year. And, once again, there was more press coverage, with articles in local newspapers.

- Each year our models grow in sophistication and whether or not a March Madness student ever claims the $10,000 ESPN prize, each student can claim mastery of essential hands-on modeling, programming, and communication skills, which have helped land some coveted job offers. In fact, Neil Goodson even heard the comment, "Hey, you're that March Madness ranking guy from NPR, aren't you?" on a job interview.

By The Numbers —

$9, 223, 372, 036, 854, 775, 808 = $ # of ways to fill out a 64-team March Madness bracket
after all teams have been seeded.

$147, 573, 952, 589, 676, 412, 928 = $ # of ways to fill out a 68-team March Madness bracket.

—It's simple arithmetic: compute 2^{63} and 2^{67}.

Chapter Thirteen

"What If . . ." Scenarios and Sensitivity

It is common for coaches and fans to conduct "what if" analysis at certain crucial points in a season. For example, with just one game left in the season, a college football coach may wonder what happens to his team's ranking if his team wins by β points in this final contest. Of course, one way to answer the coach's question is to compute the ranking prior to the game of interest, then assume the coach's desired outcome for this final game and recompute the ranking. However, for very large (think web-sized) applications, a full recomputation of the ranking for each "what if" scenario is unreasonable or even impossible. And besides, full recomputation is akin to brute force and just isn't very elegant to a mathematician. So, in this chapter, we use the mathematical tool of perturbation analysis to precisely answer such questions with minimal additional computation.

The Impact of a Rank-One Update

In this section, we use tools from perturbation analysis to quantify the effect that a small change in the season's data can have on Colley rankings. With a season in progress, at some point, a coach begins to speculate on his team's playoff chances. At the current time in the season, we compute the Colley rating. Then we assume the coach's desired outcome in his next match and recompute the rating vector. The coach's speculation about his next game is represented as a perturbation to the coefficient matrix of the Colley system. We will study a specific type of perturbation, called a rank-one update. Rank-one updates are the simplest type of update to a matrix, and thus, make a nice starting point for our analysis. The formulas derived in this section only need to be modified slightly for updates of higher rank. In summary, we will compare the pre-perturbation rating to the post-perturbation rating. Finally we note that similar results can be derived for most other methods, yet to avoid redundancy we use just one method, the Colley method, to make our main points regarding "what if" analysis.

The tools of perturbation analysis enable us to precisely examine the effect of the coach's speculative scenario. Let $\mathbf{C}$ and $\mathbf{b}$ be the Colley matrix and right-hand side vector before the perturbation and let $\tilde{\mathbf{C}}$ and $\tilde{\mathbf{b}}$ represent the Colley matrix and right-hand side vector after the perturbation. Suppose the coach wonders what happens when his team, team j, beats team i in the upcoming game. Then the perturbed matrix $\tilde{\mathbf{C}}$ is related to the pre-perturbation matrix $\mathbf{C}$ by the formula $\tilde{\mathbf{C}} = \mathbf{C} + (\mathbf{e}_i - \mathbf{e}_j)(\mathbf{e}_i - \mathbf{e}_j)^T$. Using the Sherman-Morrison rank-one update formula [54] and some M-matrix properties of the Colley matrix, we can express $\tilde{\mathbf{r}}$ in terms of $\mathbf{C}$. Specifically,

$$\tilde{\mathbf{r}} = \tilde{\mathbf{C}}^{-1}\mathbf{b} \tag{13.1}$$
$$= (\mathbf{C} + (\mathbf{e}_i - \mathbf{e}_j)(\mathbf{e}_i - \mathbf{e}_j)^T)^{-1}\tilde{\mathbf{b}}$$
$$= \mathbf{r} - \left(\frac{r_i - r_j}{1 + [C^{-1}]_{ii} - 2[C^{-1}]_{ij} + [C^{-1}]_{jj}}\right)\mathbf{C}^{-1}(\mathbf{e}_i - \mathbf{e}_j).$$

We draw three conclusions from the above equation.

- A single additional game can affect the ratings of every team. This is because the ith and j^{th} columns of $\mathbf{C}^{-1}$ are almost always completely dense.

- Suppose, before the perturbation, that team i was ranked above team j so that $r_i - r_j > 0$. Then, as expected, with the additional event of an upset of team j over team i, the rating of team i, $\tilde{r}_i$, drops while the rating of team j, $\tilde{r}_j$, rises. This is a consequence of the diagonal dominance and nonnegativity of $\mathbf{C}^{-1}$.

- If $\mathbf{C}^{-1}$ is available, then we can just pick a few elements from it in order to form $\tilde{\mathbf{r}}$. With $\tilde{\mathbf{r}}$ in hand, we know precisely how each team's rating changes as a result of the perturbation, and thus, can answer any "what if" question about any pair of teams.

To consider multiple "what if" scenarios simultaneously, a similar analysis is conducted with rank-k updates instead of the simpler rank-one update.

Sensitivity

"What if" scenarios are actually related to the mathematical concept of sensitivity. In mathematics, we call a method sensitive if small changes in the input data create large changes in the output data. We prefer insensitive or robust methods. A sensitive ranking method may give erratic rankings after an upset. This was precisely the conclusion in [21], where a tractable but practically unlikely "perfect season" (a season that contains no upsets) is considered. In other words, in a perfect season some team will go undefeated, followed by a team that loses only to the undefeated team yet beats every other team, while a third team loses to these two teams but beats all remaining teams, and so on. If the data for a season follows this structure, then, of course, the ranking of the teams is clear. While perfect seasons are extremely rare, they create data that can be mathematically analyzed with closed form expressions [21]. In many ways, the perfect season is the most well-behaved of all input seasons. If a ranking method shows signs of sensitivity on this season, then its sensitivity problems will only be exacerbated in the more realistic case of imperfect seasons.

The study in [21] analyzed the sensitivity of the Colley, Massey, and Markov ranking methods applied to the perfect season, and the results revealed good news for the Colley and Massey methods. Both were found to be insensitive to small rank-one changes to the system, with the Massey being particularly insensitive. On the other hand, the Markov method was shown to be sensitive to small rank-one changes to the system. One particular event from the 2008 NFL season was pointed to that highlighted Markov's sensitivity on an imperfect season. This event occurred when last place Cleveland at rank 32 upset the 17-ranked NY Giants on 10/13/2008 during the Monday Night game. Comparisons of the

Colley, Massey, and Markov rankings both before and after the Monday Night event are shown below.

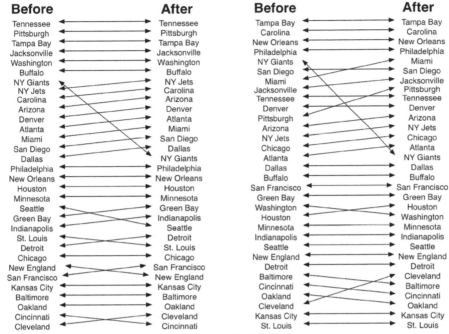

Colley Method

Before → **After**

Before	After
Tennessee	Tennessee
Pittsburgh	Pittsburgh
Tampa Bay	Tampa Bay
Jacksonville	Jacksonville
Washington	Washington
Buffalo	Buffalo
NY Giants	NY Jets
NY Jets	Carolina
Carolina	Arizona
Arizona	Denver
Denver	Atlanta
Atlanta	Miami
Miami	San Diego
San Diego	Dallas
Dallas	NY Giants
Philadelphia	Philadelphia
New Orleans	New Orleans
Houston	Houston
Minnesota	Minnesota
Seattle	Green Bay
Green Bay	Indianapolis
Indianapolis	Seattle
St. Louis	Detroit
Detroit	St. Louis
Chicago	Chicago
New England	San Francisco
San Francisco	New England
Kansas City	Kansas City
Baltimore	Baltimore
Oakland	Oakland
Cincinnati	Cleveland
Cleveland	Cincinnati

Massey Method

Before → **After**

Before	After
Tampa Bay	Tampa Bay
Carolina	Carolina
New Orleans	New Orleans
Philadelphia	Philadelphia
NY Giants	Miami
San Diego	San Diego
Miami	Jacksonville
Jacksonville	Pittsburgh
Tennessee	Tennessee
Denver	Denver
Pittsburgh	Arizona
Arizona	NY Jets
NY Jets	Chicago
Chicago	Atlanta
Atlanta	NY Giants
Dallas	Dallas
Buffalo	Buffalo
San Francisco	San Francisco
Green Bay	Green Bay
Washington	Houston
Houston	Washington
Minnesota	Minnesota
Indianapolis	Indianapolis
Seattle	Seattle
New England	New England
Detroit	Detroit
Baltimore	Cleveland
Cincinnati	Baltimore
Oakland	Cincinnati
Cleveland	Oakland
Kansas City	Kansas City
St. Louis	St. Louis

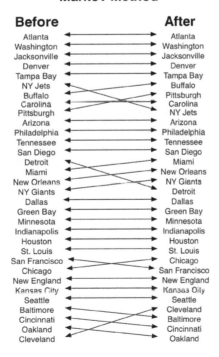

Markov Method

Before → **After**

Before	After
Atlanta	Atlanta
Washington	Washington
Jacksonville	Jacksonville
Denver	Denver
Tampa Bay	Tampa Bay
NY Jets	Buffalo
Buffalo	Pittsburgh
Carolina	Carolina
Pittsburgh	NY Jets
Arizona	Arizona
Philadelphia	Philadelphia
Tennessee	Tennessee
San Diego	San Diego
Detroit	Miami
Miami	New Orleans
New Orleans	NY Giants
NY Giants	Detroit
Dallas	Dallas
Green Bay	Green Bay
Minnesota	Minnesota
Indianapolis	Indianapolis
Houston	Houston
St. Louis	St. Louis
San Francisco	Chicago
Chicago	San Francisco
New England	New England
Kansas City	Kansas City
Seattle	Seattle
Baltimore	Cleveland
Cincinnati	Baltimore
Oakland	Cincinnati
Cleveland	Oakland

As expected, with the insensitive Colley and highly insensitive Massey methods, Cleveland rose and the Giants fell a few positions in rank as a result of the upset. However, the Markov method had the Giants moving up one position despite its loss to last place Cleveland. The Markov method's display of such odd behavior on this event from an imperfect NFL season is consistent with the theoretical results that had established for the perfect season. More specifically, the conclusion in [21] was was that the Markov method is sensitive to events that involve teams in the tail of the rating distribution—Cleveland was the caboose of this tail in our NFL example. And because Markov ratings are known to follow a power law distribution [58], there are many teams in the tail. One might think that the method's sensitivity would make it suspect, yet the Markov method was at the heart of Google's original method for ranking webpages.

ASIDE: Upsets and the Markov Method's Sensitivity

While in mathematics sensitivity is generally perceived negatively, at least some parties can benefit from the Markov method's sensitivity. If the Markov method is being used to rank teams, then coaches and players of teams in the tail of the rating can attach a disproportionate amount of hope to an upset. It is possible for low-ranked teams to catapult several, and sometimes many, positions up the ranked list if they pull off an upset, beating a higher ranked team even by just one point. And the bigger the upset (i.e., the higher ranked the opponent), the greater the potential for rising in rank.

By The Numbers —

— The biggest upsets of all time.

1.   "The Miracle On Ice" — U.S. beats the Soviets Olympic ice hockey team, 1980.
2.   New York Jets defeat Baltimore Colts, Super Bowl III, 1969.
3.   Villanova tops Georgetown to win 1985 NCAA men's basketball championship.
4.   Buster Douglas KOs Mike Tyson for the heavyweight championship, 1990.
5.   Man O' War loses his only race to the 100-to-1 long shot named Upset, 1919.
6.   Denver Nuggets eliminate Seattle SuperSonics in the 1994 NBA playoffs.
7.   Jack Fleck beats Ben Hogan at the 1955 U.S. Open.
8.   New York Mets defeat Baltimore Orioles for the 1969 World Series.
9.   Rulon Gardner overpowers Alexander Karelin, Greco-Roman wrestling gold, 2000.
10.   N.C. State beats Houston to win the 1983 NCAA men's basketball championship.

— espn.go.com/page2/s/list/topupsets/010525.html

Chapter Fourteen
Rank Aggregation–Part 1

The dictum "the whole is greater than the sum of its parts" is the idea behind rank aggregation, which is the focus of this chapter and the next. The aim is to somehow merge several ranked lists in order build a single new superior ranked list.

The need for aggregating several ranked lists into one "super" list is common and has many applications. Consider, for instance, meta-search engines such as Excite, Hotbot, and Clusty that combine the ranked results from the major search engines into one (hopefully) superior ranked list. Figure 14.1 gives a pictorial representation of the concept of rank aggregation. Here $l_1, l_2, \ldots, l_k$ represent k individual ranked lists that are somehow

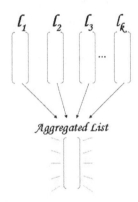

Figure 14.1 Rank aggregation of k ranked lists

combined, using mathematical techniques that will be explained in this chapter, to create one robust aggregated list. This process is called *rank aggregation* and has been studied extensively in [28], [25], and [24]. In fact, we have already encountered rank aggregation on page 82. There we needed to combine just two ranked lists, the offensive ranked list created by the OD rating vector $\mathbf{o}$ with the defensive ranked list created by the OD rating vector $\mathbf{d}$. In that case, we used the simple rule $\mathbf{r} = \mathbf{o}/\mathbf{d}$ to create an aggregated rating vector $\mathbf{r}$, which was then used to create the "robust" ranked list. In this chapter, we introduce several existing rank-aggregation methods that are more general in the sense that they can be used to combine any number of ranked lists. Then on pages 167 and 172 we introduce our contributions (two new algorithms) to the field of rank aggregation. However, before jumping into the methods we return once again to the topic that opened this book, political voting theory.

Arrow's Criteria Revisited

This book began with a chapter on Arrow's Impossibility Theorem [5], which states the limitations inherent in all voting systems. In summary, Arrow's theorem says that no voting system can ever simultaneously satisfy the four criteria of an unrestricted domain, independence of irrelevant alternatives, the Pareto principle, and non-dictatorship. At least one of these criterion must be sacrificed. In this section we consider just how concerned we are with each criterion, especially given the particulars of the sports ranking problem.

While Chapter 1 pointed out the similarity between the problem of ranking candidates in a political system and the problem of ranking teams or individuals in sporting events, there are actually some differences between the two problems that are worth exploring in greater detail.

The context for Arrow's theorem, i.e., political voting, differs from the sports ranking problem in two important respects. First, the political arena is mainly concerned with plurality voting. Creating ranked lists for all candidates is simply unnecessary because only the top ranked candidate has something to gain, whereas in sporting events such as volleyball, tennis, and NASCAR, first, second, third, and further place finishers typically receive points or cash prizes. Second, while political voting and sports ranking both have aspects of rank aggregation, the parameters in each case are quite different. Consider Figure 14.2, which shows the major differences in the data input and processing between the sports and political ranking problems.

Even if political elections asked voters to rank all candidates, there are relatively few candidates that run for most offices, while the pool of voters is huge. The situation is reversed in sports ranking, where typically only a handful of voters (ranking systems) rank hundreds of candidates (teams). And these two differences between the ranking systems, sports and politics, cause us to re-examine the justification for demanding that Arrow's criteria hold.

Arrow's first criterion requires that a voting system have an unrestricted domain, which means that every voter should be able to rank the candidates in any arrangement he or she desires. For instance, if the voting system required that an incumbent must be ranked in the top five, then Arrow's first criterion would not be satisfied. In the sports ranking context, we agree with Arrow that an unrestricted domain is a desirable property.

On the other hand, Arrow's second criterion, the independence of irrelevant alternatives, is probably the most controversial of Arrow's demands on a voting system. It states that relative rankings within subsets of candidates should be maintained when expanding back to the whole set. Many voting theorists have debated the notion that society follows this axiom naturally, and the U.S. presidential election of 2000 highlighted for some the importance of this criterion. This dramatic election was so close that a winner could not be determined until the votes from the final submitting state of Florida were tallied. And even then recounts were issued as controversies concerning miscounts, chads, and the like surfaced. Many analysts speculated that Al Gore would have beaten George W. Bush in a head-to-head election in Florida if Ralph Nader had not taken some votes away from Gore. Votes cast for Nader allowed Bush to win what was essentially a best-out-of-three competition. In this case, Gore is what is known as the *Condorcet winner* and this situation is commonly referred to as the "spoiler effect" in the social choice literature. If

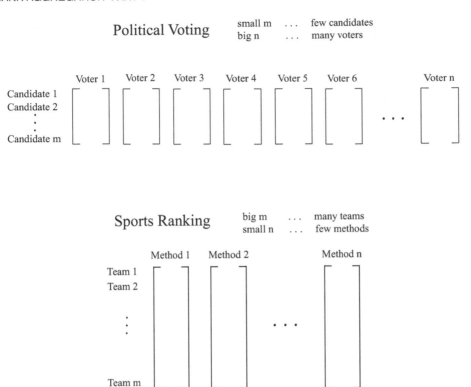

Figure 14.2 Differences in aggregating ranked lists in political voting vs. sports ranking

the 2000 election had met Arrow's independence criterion, then the rankings in the subset $\{Gore, Bush\}$ should not have been affected by the advent of the third candidate in the expanded set $\{Gore, Bush, Nader\}$. The 2000 election showed that Condorcet winners can lose if a voting system does not satisfy Arrow's second criterion.

Since politics and sports are two quite different arenas, we now debate the relevancy of independence of irrelevant alternatives to our application of ranking sports teams. Nearly all sports rating methods use pair-wise matchups of teams. Note that "teams" can also be read as "candidates" to continue the comparison with political voting systems. We have already mentioned that for most sports, especially collegiate ones, a team does not face every other team during the season. There are just too many collegiate teams and not enough time to make this a realistic possibility. (Those scholar-athletes have to attend classes sometime.) Nevertheless, despite the fact that VT may never play USC during the season, we can still infer an indirect relationship between the two teams based upon inter-conference play. As a result, opponents of opponents can reveal once-removed connections between teams. That is, USC may have beaten UNC, who played and defeated VT. Thus, by the transitive property, one could argue that USC ought to be ranked above VT. Consequently, it can be argued that removing a team from a set of rankings *should* affect the overall rankings. And this statement leads us to reject Arrow's second criterion for a rating system.

Yet another argument against the need for the independence of irrelevant alternatives can be made with the simple example of ranking wrestlers. It is common in many sports including wrestling, that j has an advantage over i in head-to-head matchups due to, for example, clashes in style. Yet j may be consistently defeated by most opponents while i performs very well against most opponents. As a result, wrestler i will perform very well in tournaments where he does not face j early on. When expanding to the set of all wrestlers in a tournament, i will almost always be ranked ahead of j after the tournament given that i and j did not wrestle in the first round. This is a violation of the independence criterion, which would require that j remain ranked ahead of i after expanding to the larger set of wrestlers. In summary, Arrow's second criterion of the independence of irrelevant alternatives does not seem nearly as useful for the sports ranking problem as it does for the political voting problem.

However, before we completely write off Arrow's second criterion in the sports context, let us, for a moment, play the devil's advocate and try to imagine a sporting scenario in which independence of alternatives is valuable. We use our running example to create just such a scenario. Recall from Table 1.1 that Miami, VT, UNC, and UVA all defeated Duke. A question that relates to the independence of irrelevant alternatives is: should the removal of Duke from Table 1.1 really affect the rankings? In other words, after the removal of Duke, should Miami's rating be boosted more than UVA's rating because Miami beat Duke by twenty more points than UVA beat Duke? Arrow would answer "No", while the previous paragraph argued, "Yes, in some cases." These are tricky questions. It is possible that, when facing Duke, a weak opponent, the UVA coach put in his second-string players, while Miami's coach did not. Thus, perhaps a win over Duke should really carry very little weight, and thus, if Duke were removed from the set of candidates, then the rankings should be unaffected, which creates an argument in support of Arrow's second criterion. Systems that incorporate game scores must reconcile the benefits of using point data with the problem of strong teams purposely running up scores on weak teams. This "running up the score" issue appears to be the primary reason why the BCS prefers models that do not involve game scores, and instead only use win-loss records. Recall the bias-free notion embodied by the Colley method of Chapter 3. This scenario also highlights the fact that information regarding personnel changes or injuries is information that is not factored into any of the models in this book.

The above paragraphs present two ways of considering the independence criterion. In the first case, we presented an argument against it, and in the second case, an argument for it. Yet another way to examine this independence criterion is to take a mathematician's perspective. Mathematicians often consider a model's sensitivity to either small changes in the data or the prevalence of outliers. If a system gives stable results despite small changes in the input data, then the system is called robust, a desirable property. On the other hand, systems that are highly sensitive are undesirable. Consequently, it is natural to connect Arrow's concept of independence of irrelevant alternatives with the mathematical notions of sensitivity and robustness. That is, a robust ranking method would satisfy Arrow's second criterion.

There is much less debate over Arrow's third criterion of the Pareto principle, which states that if all voters choose candidate A over B, then a proper voting system should maintain this order. This is clearly a desirable criterion to satisfy in both the political voting context and the sports ranking context. For example, if all methods rank Miami ahead of Duke, then an aggregated ranking of these methods should have Miami ranked ahead of Duke.

Arrow's fourth and final criterion is non-dictatorship. While it seems obvious that no voter should have more weight than another in the political setting, this is not so obvious in the sports ranking context for two reasons. First, some of the aggregation techniques discussed in the next section require a "tie-breaking" strategy between lists of close proximity. In this case, usually one list is given priority over another by the aggregation function. Thus, the list given priority acts as a dictator or at least a partial dictator. In a second scenario, there may be a justifiable reason to have a (partial) dictator. For example, sports analysts may want to have rankings generated by coaches and experts count more (or less) than rankings generated by computer models. Such wishes can be accommodated if we are willing to break Arrow's fourth criterion. As a result, it appears that Arrow's final criterion is not as useful in the sports ranking context as it is in the political voting context.

Rank-Aggregation Methods

The idea of rank aggregation is not new. As the previous section implied, it has been studied for decades in the context of political voting theory. However, the field of web search brought renewed interest in rank aggregation. With the dramatic increase in the past decade of spamming techniques that aim to deceive search engines and achieve unjustly high rankings for client webpages, search engines needed to devise anti-spam counter methods. And this is precisely why and when rank-aggregation methods received renewed attention. For example, meta-search engines are based on the idea that aggregating the ranked lists produced by several search engines creates one super parent list that inherits the best properties of the child lists. Thus, if one child list from, say, Google contains an outlier (a page that has been successfully optimized by its owner to spam Google into giving it a very high though unjust ranking), then, after the rank aggregation, the effect of this outlier, this anomaly, should be mitigated. As a result, rank aggregation tends to act as a "smoother," meaning that it removes any anomalous bumps or outliers in the rankings of individual lists and smoothes the results. Another property of a rank-aggregated list is that it is only as good (or as bad) as the lists from which it is built. That is, one cannot expect to create an extremely accurate rank-aggregated list given a set of poor input lists. In this section, we present four methods of rank aggregation. The first two are historically common ones, while the last two are our new contributions to the field of rank aggregation.

ASIDE: Web Spam and Meta-search Engines

This aside emphasizes the prevalence and pervasiveness of web spam by discussing two popular spamming techniques, Google bombs and link farms [49]. By the end, the need for spam-resistant rating methods will be clear. Fortunately, meta-search engines, which use rank-aggregation methods, such as the ones discussed in this chapter, have been successful at deterring spam.

A Google bomb is a spamming technique that exploits the high importance that Google places on anchor text. Perhaps the most famous Google bomb was aimed at G. W. Bush. Figure 14.3 shows the mechanism behind the bomb. The effectiveness of a Google bomb

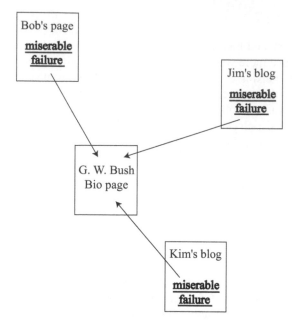

Figure 14.3 How a Google bomb works

is tied to Google's belief that the anchor text associated with a hyperlink serves as a short summary or synopsis of the webpage it points to. If several pages (in the example of Figure 14.3, Bob's page, Jim's blog, and Kim's blog) all point to Bush's White House biography page and carry the same anchor text of "miserable failure," then Google assumes that Bush's page must be about this topic. When enough spammers collude, this spamming strategy causes the target page, i.e., Bush's page, to rank highly for the search phrase, i.e., "miserable failure." When Google went public in 2004, their IPO documentation cited such spamming as a risk that investors must consider. The rank-aggregation techniques of this chapter offer one remedy to Google bombs. A meta-search engine that aggregates results from several major engines would find Google's results for spammed phrases such as "miserable failure" to be inconsistent with the results from other search engines. The effect of these outliers would be mitigated by the rank-aggregation processes described in the next few sections.

A second popular spamming technique is the link farm. A link farm affects any search engine that incorporates link-based analysis into its ranking of webpages. Most link analysis algorithms, particularly Google and its PageRank algorithm, reward pages that receive inlinks from other high ranking pages. A link farm creates high ranking pages on general topics, then shares this PageRank with client pages. For example, consider Figure 14.4. A search engine optimization (SEO) firm may create a densely connected group of pages on topics such as former U.S. presidents. These pages contain valid relevant information on these topics and thus receive decent rankings from a search engine. When a customer contacts a SEO firm hoping to boost the rank of its page, a link is simply made from the link farm to the client's page. Thus, in effect, the link farm was artificially created in order to share its PageRank with clients. Link farms are hard for search engines to detect and thus are a troublesome spamming technique. Just as in the case of Google bombs, one possible way to combat link farms is to employ the rank-aggregation techniques discussed in the next few sections.

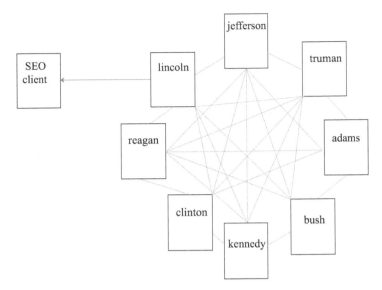

Figure 14.4 How a link farm works

Borda Count

A common rank-aggregation technique is Borda's method [12], which dates back to 1770 and Jean-Charles de Borda. Borda was trying to aggregate ranked lists of candidates from a political election. For each ranked list, each candidate receives a score that is equal to the number of candidates he or she outranks. The scores from each list are summed for each candidate to create a single number, which is called the Borda count. The candidates are ranked in descending order based on their Borda count. The BCS ranking method discussed in the aside on page 17 is popular example of how Borda counts are applied.

Though Borda's method is very simple, an example provides the best, most concise explanation. And so we refer once again to our running 5-team example. Choosing the Borda count as the method for aggregating the three ranked lists from OD, Massey, and Colley, Table 14.1 lists the items needed for this computation. The number 4 listed next to VT in the OD column means that there were four teams that VT outranked in the OD list. By summing the rows for the three methods, we have a Borda count aggregation of the three lists.[1]

This example demonstrates the simplicity and computational ease associated with the Borda method for rank aggregation. In addition, the Borda count method of rank aggregation can be adapted to handle ties. For example, suppose a voter ranked the five teams from best to worst as {Miami, VT, UNC/UVA, Duke}, where the slash (/) indicates a tie. In this case, the Borda scores for this list are

$$
\begin{array}{c}
\text{Duke} \\
\text{Miami} \\
\text{UNC} \\
\text{UVA} \\
\text{VT}
\end{array}
\begin{pmatrix}
0.0 \\
4.0 \\
1.5 \\
1.5 \\
3.0
\end{pmatrix}.
$$

---

[1]We thank Luke Ingram for the creation of this data and table.

Table 14.1 Borda count method for aggregating rankings from OD, Massey, and Colley for 5-team
example

	OD ($r = o/d$)	Massey	Colley	Borda Count	Borda Rank
Duke	0	0	0	0	5th
Miami	3	4	4	11	1st
UNC	1	1	2	4	4th
UVA	2	2	1	5	3rd
VT	4	3	3	10	2nd

Notice that teams that tie split the Borda scores of the fixed rank positions. Thus, the Borda count method can handle input lists with ties. In addition, though unlikely, the Borda count method could produce an output aggregate list that contains ties.

Unfortunately, the Borda method has one serious drawback in that it is easily manipulated. That is, dishonest voters can purposefully collude to skew the results in their favor. Textbooks, such as [16], provide classic examples demonstrating the manipulability of the Borda count.

ASIDE: Llull's Count

This chapter presents the mathematical ideas of the Borda count and the Condorcet winner, which both take the namesake of their inventors. However, recently discovered manuscripts reveal that these mathematical ideas really ought to be called Llull counts and Llull winners after Ramon Llull (1232-1315) , the extremely prolific Catalan writer, philosopher, theologian, astrologer, and missionary. Llull has over 250 written documents to his name. In 2001, with the discovery of three lost manuscripts, *Ars notandi, Ars eleccionis* and *Alia ars eleccionis*, historians discovered that they could add three more works to that tally and also add mathematician to Llull's list of occupations and titles. Since Llull's  work on election theory predated Borda's and Condorcet's independent discoveries by several centuries, he has since been acknowledged as the first to discover the Borda count method described in this section and the Condorcet winner discussed on page 160.

Average Rank

Another very simple rank-aggregation technique is average rank. In this case, the integers representing rank in several rank-ordered lists are averaged to create a rank-aggregated list. Again, an example makes this quite clear. Table 14.2 shows the average rank for the three ranked lists created by the OD, Massey, and Colley methods.

This example hides one drawback common to average rank—the frequent occurrence of ties in the average rating score. Let l_1 and l_2 be two *full ranked lists*. A full ranked list contains an arrangement of all elements in the set. Suppose team i is ranked first in l_1, but third in l_2, and team j is ranked second in both lists. In this case, average rank produces

Table 14.2 Average rank method of aggregating rankings from OD, Massey, and Colley for the 5-team example

	OD ($r = o/d$)	Massey	Colley	Average Rating	Average Rank
Duke	5th	5th	5th	5	5th
Miami	2nd	1st	1st	$1.\overline{3}$	1st
UNC	4th	4th	3rd	$3.\overline{6}$	4th
UVA	3rd	3rd	4th	$3.\overline{3}$	3rd
VT	1st	2nd	2nd	$1.\overline{6}$	2nd

a tie for second place. To further complicate matters average rank can produce multi-way ties at multiple positions.

There are several strategies for breaking ties such as the one mentioned in the previous paragraph. We describe two. The first strategy makes use of past data to break ties. If teams i and j played each other, then the winner of that matchup (or the "best of" winner of multiple matchups) should be ranked ahead of the losing team. Retroactively using game outcomes to determine a tie-breaking team is more difficult when averaging more than two lists because several difficulties can arise. For example, suppose that teams i, j, and k tie for a rank position when three or more ranked lists are averaged. It is possible that the data used to create the lists contains the following outcomes: i defeated j, j defeated k, and k defeated i. This type of circular tie is a frequently cited difficulty with rankings created by pair-wise matchups. A typical solution is to assign one list as the tie-breaker list, which is a strategy described in the next paragraph.

If the teams involved in the tie never played, then the second tie-breaking strategy is to assign one ranked list as the "tie-breaking list," which works well in the case of pair-wise ties. For example, suppose a tie scenario occurs in which teams i and j both vie for second position in the aggregated list. Yet suppose that the first ranked list has been chosen as the superior, and thus, tie-breaking list. Then this means that team i wins the second place position in the aggregated list and team j fills in the third position. The choice of tie-breaker list need not be arbitrary either, particularly if we have a method for determining which ranked list is better. Measures for determining the quality of a ranked list are discussed in Chapter 16.

We close this section with some observations regarding the average rank method of aggregation. First, note that average rank can be applied only if all lists are *full* (i.e., the intersection of the elements in all lists contains all teams). Second, the average rank method of aggregation can produce an aggregated list that does not contain ranks. This was clear from Table 14.2. Of course, the solution there was easy. The result of averaging ranked lists produces an average rating vector, which can then be transformed into a ranking vector.

Simulated Game Data

In this section we present a new rank-aggregation method that is born from a simple interpretation of a ranked list. See Figure 14.5. The interpretation is this: *if team A appears above team B in a ranked list, then in a matchup between these two teams A ought to beat*

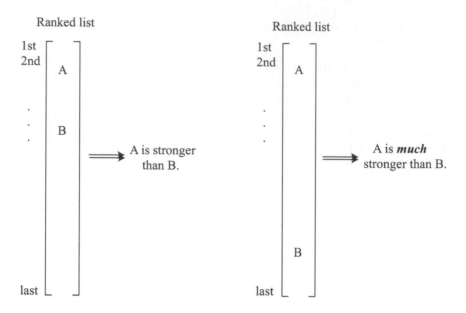

Figure 14.5 Interpretation of relative position of items in a ranked list

B. Further, if team A appears at the top of the ranked list and plays team B, which appears at the bottom of the list, then we expect A to beat B by a large margin of victory. The observation that a ranked list gives implicit information about future game outcomes seems rather natural. And yet the clever part, in terms of the mathematical modeling, is that we use ranked lists to generate so-called *simulated game data*, which we eventually use to aggregate the results of several lists into one list. In fact, each ranked list of length n provides data for $\binom{n}{2} = n(n-1)/2$ simulated games. Let's return again to our 5-team example and the three ranked lists of OD, Massey, and Colley, which appear below.

	OD		*Massey*		*Colley*
1st	VT	1st	Miami	1st	Miami
2nd	Miami	2nd	VT	2nd	VT
3rd	UVA	3rd	UVA	3rd	UNC
4th	UNC	4th	UNC	4th	UVA
5th	Duke	5th	Duke	5th	Duke

Starting with the OD ranking vector, we can form $\binom{5}{2} = 10$ pair-wise matchups between these five teams. If VT played Miami, we'd expect VT to win, based on the information in the OD vector. Further, we can assign a margin of victory to this simulated game, if we simply assume that margins are related to the difference in the ranked positions of the two teams. This is where we, as modelers, have lots of wiggle room for creative ideas. Here we present one elementary idea that works well. We assign a margin of victory of one point for each difference in the rank positions. Thus, according to the OD vector, VT would beat Miami by one point, UVA by 2 points, UNC by 3 points, and Duke by 4 points. This point score data is useful later in the rank-aggregation step. After we have generated simulated game data for the OD ranking vector, we move onto the next ranking vector, and so on, until we have accumulated a mound of simulated game data. Note that any number of ranked lists can be used to build the complete set of simulated data. In the above example, with three ranked lists of five teams each, we generate $3\binom{5}{2} = 30$ simulated games. We also note that, *unlike most other aggregation methods, the lists need not be full*. Some lists may be full, some partial, and the simulated game data can be built nonetheless.

Once the simulated data has been generated completely, the aggregation actually occurs in the next step, which we describe now. At this step, any ranking method is applied to the simulated game data. This ranking method, which we call the *combiner method*, could be an entirely different method. For instance, a method not used to create the input ranking vectors could be applied. Or on the other hand, a favored method from those used as input can be applied as the combiner method. In this case, that favored popular method, which has been used as both an input method and a combiner method, becomes in some sense a partial dictator, using Arrow's language.

Table 14.3 shows the simulated game data built from the OD, Massey, and Colley ranking vectors. Notice that we have set the losing team's score to 18. We could have

Table 14.3 Simulated game data

	Duke	Miami	UNC	UVA	VT
Duke		18-21	18-20	18-20	18-22
		18-22	18-19	18-20	18-21
		18-22	18-19	18-19	18-21
Miami	21-18		20-18	19-18	18-19
	22-18		21-18	20-18	19-18
	22-18		20-18	21-18	19-18
UNC	19-18	18-20		18-19	18-19
	19-18	18-21		18-19	18-20
	20-18	18-20		19-18	18-21
UVA	20-18	18-19	19-18		18-20
	20-18	18-20	19-18		19-20
	19-18	18-21	18-19		18-20
VT	22-18	19-18	21-18	20-18	
	21-18	18-19	20-18	20-19	
	21-18	18-19	19-18	20-18	

chosen 0, 10, or any other arbitrary number (more wiggle room for experimentalists), yet since 18 was the average losing team's score for the 2005 season, this is reasonable. Notice also that each cell of the 5 × 5 table contains three game outcomes, one for each of the three ranking vectors used as input. In particular, the 18-21 entry in the (Duke, Miami) cell indicates that in the OD ranking vector Duke appeared three rank positions below Miami and thus loses in a simulated matchup by three points. The 18-22 entry in that same cell comes from the Massey ranking vector, which had Duke four rank positions below Miami.

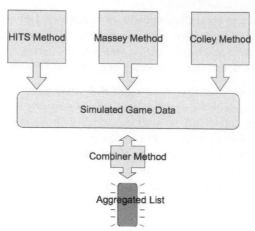

Simulated data method for rank aggregation

The diagram shown above gives a pictorial summary of the simulated data approach to rank aggregation. This approach is then used to aggregate the data from the OD, Massey, and Colley vectors for the small 5-team example. Table 14.4 shows that the aggregated results can differ slightly depending on the combiner method used.

Table 14.4 Simulated data method of rank aggregation using OD, Massey, and Colley as the combiner method

	Combiner Method		
	OD ($r = o/d$)	Massey	Colley
Duke	5th	5th	5th
Miami	1st	1st	1st
UNC	3rd	4th	4th
UVA	4th	3rd	3rd
VT	2nd	2nd	2nd

We point out four properties of the simulated data approach to rank aggregation.

• The rank-aggregated list is only as good (or as bad) as the input lists.

• The combiner method acts as a "smoother" in that it minimizes the effect of outliers, which are lists containing anomalies that seem inconsistent with the rankings in other lists. This comment whispers once again of Arrow's second criterion, the independence of irrelevant alternatives, and the related issues of robustness and sensitivity. As an example of the impact of the combiner method, consider Table 14.4. To test for independence, we removed Duke from the data set, which forces a swap in the rankings between UNC and UVA when the Massey and Colley methods are used as the combiner methods. Compare Table 14.5 with Table 14.4. In contrast, using OD as the combiner method, creates a consistent ranking that is unaffected by the removal of Duke, the weakest team. Consequently, for this very small example, we say that the OD rank-aggregation method is robust.

Table 14.5 Simulated data method of rank aggregation and its sensitivity to the removal of Duke

Combiner	OD ($r = o/d$)	Massey	Colley
Miami	1st	1st	1st
UNC	3rd	**3rd**	**3rd**
UVA	4th	**4th**	**4th**
VT	2nd	2nd	2nd

- Rank-aggregated lists satisfy Arrow's third criterion, the Pareto principle, so long as the input lists satisfy the Pareto principle.

- Though it is a small example, Table 14.4 shows that the combiner method can have an effect on the aggregated list. In addition, if the combiner method is also one of the input methods, then this method can be called a partial dictator in the language of Arrow's fourth criterion.

ASIDE: Top 10 Books—Aggregating Human-generated Lists

Most lists of the best books of all time are human-generated. In other words, surveys are conducted and votes are tallied. Book clubs, publishers, magazines, and the like each produce their top-10 lists of the best books ever. In June 2009 *Newsweek* produced their **meta-list** of the top-100 books. That is, they gathered ten top-10 lists from sources such as *Time*, Oprah's Book Club, Wikipedia, and the New York Public Library and aggregated these lists to create their top-100 meta-list. *Newsweek* used a simple weighted voting scheme to aggregate these lists, but of course, they could have applied any of the rank-aggregation methods in this chapter (or the next) instead. For all the book-loving readers, the top-10 books of all time, according to *Newsweek's* meta-list, are shown in Table 14.6

Table 14.6 Top-10 books of all-time according to *Newsweek's* meta-list

Rank	Title	Author
1	*War and Peace*	Leo Tolstoy
2	*1984*	George Orwell
3	*Ulysses*	James Joyce
4	*Lolita*	Vladimir Nabokov
5	*The Sound and the Fury*	William Faulkner
6	*Invisible Man*	Ralph Ellison
7	*To the Lighthouse*	Virginia Woolf
8	*The Illiad and the Odyssey*	Homer
9	*Pride and Prejudice*	Jane Austen
10	*Divine Comedy*	Dante Alighieri

Graph Theory Method of Rank Aggregation

In this section we propose yet another method for aggregating several ranked lists. This method relies on graph theory. Consider the graph of Figure 14.6 below in which each team is a node in the graph. The weights are not game scores or, for that matter, any other

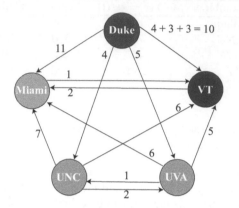

Figure 14.6 Graph for graph theory approach to rank aggregation for the 5-team example

statistical game data. Instead, the weights on the links in Figure 14.6 come from information in the ranked lists that we want to aggregate. For example, consider the following two suggestions for attaching a weight w_{ij} to the link from i to j.

$$w_{ij} = \text{number of ranked lists having } i \text{ below } j \qquad (14.1)$$

or

$$w_{ij} = \text{sum of rank differences from lists having } i \text{ below } j. \qquad (14.2)$$

The weights can be considered votes that weak teams cast for strong teams. These are just two possible ways to cast votes.

The discussion on page 176 explores an extension called *rating aggregation* that creates weights in the aggregated graph from the numerical values in the *rating* rather than the ranking vectors.

Once an approach for attaching weights to the links in the dense team graph has been selected, a favorite node ranking algorithm can be applied to this weighted graph. For instance, any of the popular node ranking algorithms from web search including PageRank, HITS, or SALSA, can be used. See [49] for more information on these algorithms.

We applied the PageRank method (a popular method for ranking nodes) [15] to the graph of Figure 14.6. This graph was built from the three ranking vectors displayed on p. 168 that were created by OD, Massey, and Colley. The links from Figure 14.6 are weighted according to Definition (14.2). For example, the link from Duke to VT has a weight of $4 + 3 + 3 = 10$ because Duke appears 4 rank positions below VT in the OD vector, 3 rank positions below VT in the Massey vector, and 3 rank positions below VT in the Colley vector. Applying the PageRank method to this graph gives a PageRank vector that creates an aggregated ranking vector of

$$
\begin{array}{cc}
\text{1st} & \left(\begin{array}{c} \text{Miami} \\ \text{VT} \\ \text{UVA} \\ \text{UNC} \\ \text{Duke} \end{array} \right), \\
\text{2nd} \\
\text{3rd} \\
\text{4th} \\
\text{5th}
\end{array}
$$

Graph theory aggregated ranking

which, by the way, matches the aggregated vector from Table 14.4 on p. 170 created from the simulated data method when Massey or Colley were used as the combiner method.

The two new rank-aggregation methods on pages 167 and 172 are much more sophisticated than the standard and existing methods discussed on pages 165 and 166. In fact, any serious rank-aggregation application, including college football's hotly contested BCS rankings, should adopt one of these more advanced methods for aggregating the results of many ranked lists. Further, the next chapter presents an even more mathematically advanced method for rank aggregation, one that guarantees that the aggregated list optimally agrees with the input lists. See Chapter 15 for more on this aggregation method.

ASIDE: Rank Aggregation and Meta-search Engines

A meta-search engine combines the results from several search engines to produce its own ranked list for a particular query. In this aside, we demonstrate how a meta-search engine could use the graph theory method of rank aggregation presented in this section. First, we gathered four top-10 ranked lists by entering the query "rank aggregation" into the four search engines of Google, Yahoo, MSN-bing, and Ask.com. The union of these four lists contains $n = 17$ distinct webpages. Applying the graph theory method results in the aggregation matrix $\mathbf{W}$ of size 17×17. The PageRank vector of this graph produces the rank-aggregated list of webpages shown on page 174 that contains the results of our meta-search engine.

To implement the simple meta-search engine, proceed as follows.

1. Get the top-k lists from your p favorite search engines.

2. Create an aggregated matrix $\mathbf{W}$ using techniques described on page 172 (or any of the rank-aggregation techniques of this chapter or the next). The dimension of $\mathbf{W}$ is $n \times n$ where n is the number of distinct webpages in the union of the p lists.

3. Compute the ranking vector $\mathbf{r}$ associated with $\mathbf{W}$.

Because k and p are typically small ($k < 200, p < 10$), $\mathbf{W}$ is small, and $\mathbf{r}$ can be computed in real-time.

[PDF] **Rank Aggregation** Methods for the Web
File Format: PDF/Adobe Acrobat - View as HTML
by C Dwork - Cited by 456 - Related articles - All 28 versions
ing the importance of **rank aggregation** for Web applications .... The ideal scenario for **rank aggregation** is when each judge ...
citeseerx.ist.psu.edu/viewdoc/download?doi=10.1.1.28.8702&rep...

[PDF] **Rank Aggregation** Revisited
File Format: PDF/Adobe Acrobat - Quick View
by C Dwork - Cited by 28 - Related articles - All 14 versions
The **rank aggregation** problem is to combine many different rank orderings on the ... In this work we revisit **rank aggregation** with an eye toward reducing ...
citeseerx.ist.psu.edu/viewdoc/download?doi=10.1.1.113.2507...

Rank Aggregation: Together We're Strong
by F Schalekamp - Cited by 3 - Related articles - All 7 versions
whereas in the case of the **rank aggregation** problem, we .... uate different algorithms for **rank aggregation** and re- lated problems. ...
www.siam.org/proceedings/alenex/.../alx09_004_schalekampf.pdf - Similar

[PDF] Efficient similarity search and classification via **rank aggregation**
File Format: PDF/Adobe Acrobat - Quick View
by R Fagin - Cited by 140 - Related articles - All 21 versions
The **rank aggregation** problem is precisely the problem of how ... and **rank aggregation**. As a simple but powerful motivating exam- ...
www.almaden.ibm.com/cs/people/fagin/sigmod03.pdf - Similar

BioMed Central | Full text | RankAggreg, an R package for weighted ...
by V Pihur - 2009 - Cited by 1 - Related articles
Two examples of **rank aggregation** using the package are given in the ..... The RankAggreg( ) function performs **rank aggregation** using either the CE algorithm ...
www.biomedcentral.com/1471-2105/10/62 - Cached - Similar

Supervised **rank aggregation**
by YT Liu - 2007 - Cited by 20 - Related articles - All 13 versions
This paper is concerned with **rank aggregation**, the task of combining the ranking results of individual rankers at meta-search. Previously, **rank aggregation** ...
portal.acm.org/citation.cfm?id=1242638

paper.dvi
Rank aggregation has been studied in many disciplines, most extensively in the context of social choice theory, where there is a rich literature dating from the latter half of the eighteenth century. We revisit **rank aggregation** with an eye toward meta-search.
citeseerx.ist.psu.edu/viewdoc/download?doi=10.1.1.113.2...

[PDF] **Rank Aggregation** for Similar Items
213k - Adobe PDF - View as html
in which the **rankings** are noisy, incomplete, or even disjoint. ... **Rank aggregation** can be thought of as the unsupervised. analog to regression, in which the goal is to find an ...
www.eecs.tufts.edu/~dsculley/papers/mergeSimilarRank.pdf

[PDF] Unsupervised **Rank Aggregation** with Distance-Based Models
File Format: PDF/Adobe Acrobat - Quick View
by A Klementiev - Cited by 5 - Related articles
One impediment to solving **rank aggregation** tasks is the high cost associated with acquiring full or ... frame unsupervised **rank aggregation** as an optimiza- ...
www.cs.jhu.edu/~aklement/publications/icml08.pdf

CiteULike: Web metasearch: **rank** vs. score based **rank aggregation**
Search all the public and authenticated articles in CiteULike. Enter a search phrase. You can also specify ... Web metasearch: **rank** vs. score based **rank aggregation** methods Export ... **aggregation** metasearch search web www...
www.citeulike.org/user/pdlug/article/1167894

Results of our simple meta-search engine for query of "rank aggregation"

A Refinement Step after Rank Aggregation

After several lists $l_1, l_2, \ldots, l_k$ have been aggregated into one list μ using one of the procedures from the section on rank aggregation on page 163, a refinement step called local Kemenization [25] can be implemented to further improve the list μ. An aggregated list μ is said to be *locally Kemeny optimal* if there exist no pair-wise swaps of items in the list that will reduce the sum of Kendall tau measures between each input list l_i and μ where $i = 1, \ldots, k$ and μ. (The Kendall tau measure τ is carefully defined on page 203). For now it is enough to know that τ measures how far apart the ranked lists are.) Thus, the local Kemenization step considers each pair in μ and asks the question: will a swap improve the aggregated Kendall measure? An example serves to explain this refinement process. Consider the three input lists created from the OD (l_1), Massey (l_2), and Colley (l_3) methods applied to the 5-team example along with the aggregated list (μ) from the simulated data aggregation technique on page 167.

	OD l_1		*Massey l_2*		*Colley l_3*		*Sim.Data μ*
1st	VT	1st	Miami	1st	Miami	1st	Miami
2nd	Miami	2nd	VT	2nd	VT	2nd	VT
3rd	UVA	3rd	UVA	3rd	UNC	3rd	UNC
4th	UNC	4th	UNC	4th	UVA	4th	UVA
5th	Duke	5th	Duke	5th	Duke	5th	Duke

The local Kemenization procedure begins with the highest ranked item (Miami) in the aggregated list μ and asks the question: does its neighboring item (VT) beat it (Miami) in the majority of the input lists? If yes, swap the two items as this creates a refined aggregated list with reduced Kendall distance. The same question is asked of each neighboring pair in the sorted aggregated list. Thus, for an aggregated list of length n, local Kemenization requires $n - 1$ checks, one for each neighboring pair.

LOCAL KEMENIZATION ON THE 5-TEAM EXAMPLE

Check #1

Question: Does the second place item (VT) beat the first place item (Miami) in the majority of the input lists?

Answer: No, VT beats Miami only once in the three input lists.

Action: None.

Check #2

Question: Does the third place item (UNC) beat the second place item (VT) in the majority of the input lists?

Answer: No, UNC never beats VT in the three input lists.

Action: None.

Check #3

Question: Does the fourth place item (UVA) beat the third place item (UNC) in the majority of the input lists?

Answer: Yes, UVA beats UNC in two out of three input lists.

Action: Swap UNC and UVA in μ.

Check #4

Question: Does the fifth place item (Duke) beat the fourth place item (UVA) in the majority of the input lists?

Answer: No, Duke never beats UVA in the three input lists.

Action: None.

Thus, the locally Kemenized aggregated list $\bar{\mu}$ is $\begin{matrix} \text{1st} \\ \text{2nd} \\ \text{3rd} \\ \text{4th} \\ \text{5th} \end{matrix} \begin{pmatrix} \text{Miami} \\ \text{VT} \\ \text{UVA} \\ \text{UNC} \\ \text{Duke} \end{pmatrix}$, which agrees with

the aggregated lists produced by the Massey combined simulated data method and the Colley combined simulated data method shown in Table 14.4 on page 170.

Rating Aggregation

So far this chapter has taken up the issue of aggregating *ranked* lists, but of course each ranked list is derived from a *rating* list, which prompts the question: can rating lists be aggregated as well? The answer is yes, though there are some hurdles to jump. For instance, one hurdle concerns scale. How do we aggregate rating vectors when each contains numerical values of such differing scale? Recall that the Colley method produces ratings that hover around .5, while Markov produces positive ratings between 0 and 1. Even more complicated, some ratings, such as Massey ratings, contain positive and negative values. Figure 14.7 shows the number line representations for the rating vectors produced by the three methods that we will consider in this section: Massey, Colley, and OD. Notice the differing scales. In order to put such varying ratings on the same scale, we invoke distances

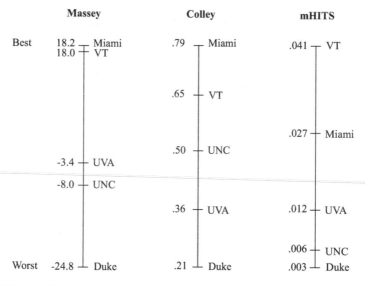

Figure 14.7 Number line representations for Massey, Colley, and OD vectors for the 5-team example

and percentages. A rating vector of length n actually contains $\binom{n}{2} = n(n-1)/2$ pair-wise comparisons between the n total teams. Consider the Massey rating vector, $\mathbf{r}_{Massey}$, below.

$$\mathbf{r}_{Massey} = \begin{matrix} \text{Duke} \\ \text{Miami} \\ \text{UNC} \\ \text{UVA} \\ \text{VT} \end{matrix} \begin{pmatrix} -24.8 \\ 18.2 \\ -8.0 \\ -3.4 \\ 18.0 \end{pmatrix}.$$

Because the rating for Miami is larger than that for Duke, the Massey method predicts Miami to beat Duke. Yet the difference between these two ratings, 43, is relatively large. Surely, this must mean something. In fact, *we like to think of these rating differences as distances*. Miami and Duke are very far apart while UNC and UVA, whose relative distance is 4.6, are close. Because it is natural to think of distances as positive, we can take differences in such a way to create an asymmetric nonnegative matrix of rating differences for each rating vector. For example, the Massey rating vector $\mathbf{r}_{Massey}$ above produces the rating distance matrix $\mathbf{R}_{Massey}$ below.

$$\mathbf{R}_{Massey} = \begin{array}{c} \\ \text{Duke} \\ \text{Miami} \\ \text{UNC} \\ \text{UVA} \\ \text{VT} \end{array} \begin{array}{ccccc} \text{Duke} & \text{Miami} & \text{UNC} & \text{UVA} & \text{VT} \\ \left(\begin{array}{ccccc} 0 & 0 & 0 & 0 & 0 \\ 43 & 0 & 26.2 & 21.6 & .2 \\ 32.8 & 0 & 0 & 0 & 0 \\ 21.4 & 0 & 4.6 & 0 & 0 \\ 42.8 & 0 & 26 & 21.4 & 0 \end{array} \right) \end{array}.$$

Computing rating distance matrices from the Colley and OD rating vectors produces

$$\mathbf{R}_{Colley} = \begin{array}{c} \\ \text{Duke} \\ \text{Miami} \\ \text{UNC} \\ \text{UVA} \\ \text{VT} \end{array} \begin{array}{ccccc} \text{Duke} & \text{Miami} & \text{UNC} & \text{UVA} & \text{VT} \\ \left(\begin{array}{ccccc} 0 & 0 & 0 & 0 & 0 \\ .58 & 0 & .29 & .43 & .14 \\ .29 & 0 & 0 & .14 & 0 \\ .15 & 0 & 0 & 0 & 0 \\ .44 & 0 & .15 & .29 & 0 \end{array} \right) \end{array}$$

and

$$\mathbf{R}_{OD} = \begin{array}{c} \\ \text{Duke} \\ \text{Miami} \\ \text{UNC} \\ \text{UVA} \\ \text{VT} \end{array} \begin{array}{ccccc} \text{Duke} & \text{Miami} & \text{UNC} & \text{UVA} & \text{VT} \\ \left(\begin{array}{ccccc} 0 & 0 & 0 & 0 & 0 \\ .024 & 0 & .021 & .015 & 0 \\ .003 & 0 & 0 & 0 & 0 \\ .009 & 0 & .006 & 0 & 0 \\ .038 & .014 & .035 & .029 & 0 \end{array} \right) \end{array}.$$

A quick glance at these $\mathbf{R}$ matrices demonstrates the problem of scale. In order to put these distances on the same scale we use the age-old trick of normalization. We divide each element of a raw $\mathbf{R}$ matrix by the sum of all distances in that matrix. The normalized matrices, denoted $\bar{\mathbf{R}}$, are below.

$$\bar{\mathbf{R}}_{Massey} = \begin{array}{c} \\ \text{Duke} \\ \text{Miami} \\ \text{UNC} \\ \text{UVA} \\ \text{VT} \end{array} \begin{array}{ccccc} \text{Duke} & \text{Miami} & \text{UNC} & \text{UVA} & \text{VT} \\ \left(\begin{array}{ccccc} 0 & 0 & 0 & 0 & 0 \\ .1792 & 0 & .1092 & .09 & .0008 \\ .1367 & 0 & 0 & 0 & 0 \\ .0892 & 0 & .0192 & 0 & 0 \\ .1783 & 0 & .1083 & .0892 & 0 \end{array} \right) \end{array},$$

$$\bar{\mathbf{R}}_{Colley} = \begin{array}{c} \\ \text{Duke} \\ \text{Miami} \\ \text{UNC} \\ \text{UVA} \\ \text{VT} \end{array} \begin{array}{ccccc} \text{Duke} & \text{Miami} & \text{UNC} & \text{UVA} & \text{VT} \\ \left(\begin{array}{ccccc} 0 & 0 & 0 & 0 & 0 \\ .2 & 0 & .1 & .1483 & .0483 \\ .1 & 0 & 0 & .0483 & 0 \\ .0517 & 0 & 0 & 0 & 0 \\ .1517 & 0 & .0517 & .1 & 0 \end{array} \right) \end{array},$$

and

$$\bar{\mathbf{R}}_{OD} = \begin{array}{c} \\ \text{Duke} \\ \text{Miami} \\ \text{UNC} \\ \text{UVA} \\ \text{VT} \end{array} \begin{pmatrix} \text{Duke} & \text{Miami} & \text{UNC} & \text{UVA} & \text{VT} \\ 0 & 0 & 0 & 0 & 0 \\ .1237 & 0 & .1082 & .0773 & 0 \\ .0155 & 0 & 0 & 0 & 0 \\ .0464 & 0 & .0309 & 0 & 0 \\ .1959 & .0722 & .1804 & .1495 & 0 \end{pmatrix}.$$

The interpretation of the .1792 element in the (Miami, Duke)-entry of the normalized $\bar{\mathbf{R}}_{Massey}$ matrix is that the distance between Miami and Duke is 17.92% of the total distance predicted between all pair-wise matchups in the Massey model. The great advantage here is that *percentages are measures that can be compared across different methods*. Further, and more to the point of this section, ratings can now be aggregated. A simple rating aggregation method takes the (possibly weighted) average of several normalized $\bar{\mathbf{R}}$ matrices. For example, averaging the normalized $\bar{\mathbf{R}}_{Massey}$, $\bar{\mathbf{R}}_{Colley}$, and $\bar{\mathbf{R}}_{OD}$ produces

$$\bar{\mathbf{R}}_{ave} = \begin{array}{c} \\ \text{Duke} \\ \text{Miami} \\ \text{UNC} \\ \text{UVA} \\ \text{VT} \end{array} \begin{pmatrix} \text{Duke} & \text{Miami} & \text{UNC} & \text{UVA} & \text{VT} \\ 0 & 0 & 0 & 0 & 0 \\ 0.1676 & 0 & 0.1058 & 0.1052 & 0.0164 \\ 0.0841 & 0 & 0 & 0.0161 & 0 \\ 0.0624 & 0 & 0.0167 & 0 & 0 \\ 0.1753 & 0.0241 & 0.1135 & 0.1129 & 0 \end{pmatrix}.$$

Producing Rating Vectors from Rating Aggregation-Matrices

Each nonzero entry of matrix $\bar{\mathbf{R}}_{ave}$ gives information about the distance between two teams, and thus, predictions about potential winners in future matchups. And with just a bit more work it may be possible to squeeze even more information out of $\bar{\mathbf{R}}_{ave}$ to produce point spread predictions, the subject of future work. For now, we discuss methods for collapsing the information in the two-dimensional matrix into a one-dimensional rating vector. We present three methods for creating a rating vector $\mathbf{r}$ from $\bar{\mathbf{R}}_{ave}$.

Method 1

The first method uses row and column sums. The row sums of $\bar{\mathbf{R}}_{ave}$ give a measure of a team's offensive output while the column sums give a measure of a team's defensive output. That is, the offensive rating vector is $\mathbf{o} = \bar{\mathbf{R}}_{ave}\mathbf{e}$ and the defensive rating vector is $\mathbf{d} = \mathbf{e}^T\bar{\mathbf{R}}_{ave}$, where $\mathbf{e}$ is the vector of all ones. Just like the OD model of Chapter 7, a high row sum corresponds to an effective offensive team and a low column sum corresponds to an effective defensive team. Thus, the overall rating vector for a team can be computed as $\mathbf{r} = \mathbf{o}/\mathbf{d}$.

Method 2

The second method for extracting a rating vector $\mathbf{r}$ from the $\bar{\mathbf{R}}_{ave}$ matrix applies the Markov method to a row-normalized version of $\bar{\mathbf{R}}_{ave}^T$. This works because the entries of $\bar{\mathbf{R}}_{ave}$ are the distances between teams, or in other words, votes that one team casts for another.

Method 3

The third and final method for extracting a rating vector $\mathbf{r}$ from the matrix $\bar{\mathbf{R}}_{ave}$ uses the dominant eigenvector of $\bar{\mathbf{R}}_{ave}$. This eigenvector actually has a name—it is the Perron vector of $\bar{\mathbf{R}}_{ave}$. This vector gets its name from the Perron theorem which tells us that the dominant eigenvector of an irreducible nonnegative matrix is guaranteed to be nonnegative. Rather than checking that $\bar{\mathbf{R}}_{ave}$ is irreducible, in practice one can force irreducibility by adding $\epsilon\,\mathbf{ee}^T$ to $\bar{\mathbf{R}}_{ave}$, where ϵ is a small positive scalar.

Table 14.7 shows the results of applying these three methods to create three rating vectors associated with $\bar{\mathbf{R}}_{ave}$.

Table 14.7 Rating vectors for three methods of rating aggregation

Team	Method 1 $r = o/d$		Method 2 Markov r		Method 3 Perron r	
Duke	0	5th	.020	5th	.27	5th
Miami	16.4	2nd	.465	2nd	.58	2nd
UNC	.4	3rd	.025	3rd	.34	3rd
UVA	.3	4th	.024	4th	.33	4th
VT	26.0	1st	.466	1st	.61	1st

ASIDE: Ranking NSF Proposals

The National Science Foundation (NSF) is America's largest funding agency for scientific work. As such, each year this governmental agency receives thousands of proposals from scientists each pitching their own research. Of course, the NSF budget, though large, is finite. As a result, decisions are continually being made as to which proposals to fund and which to reject. Most funding decisions are made with a procedure known as a panel review, which relies on the service of recognized experts in the field under consideration. For example, the NSF FODAVA (Foundations of Data and Visual Analytics) division may receive 32 proposals for the 2009 year. Approximately 20 panel reviewers (i.e., FODAVA experts who have little or no conflicts of interest with the proposers) are asked to read this collection of proposals. Because the volume of proposals makes it unrealistic for each panelist to read each proposal, division directors allocate roughly the same number of proposals to each panelist. For instance, most panelists are asked to review six proposals and some seven. Generally the directors like to have each proposal reviewed by at least three or four panelists. A panelist's review consists of written comments plus, most relevant to this book, a rating. Ratings are Excellent, Very Good, Good, Fair, and Poor. Since not every proposal is read by every panelist and since some panelists may be generous in their ratings while others are more demanding, the system faces some inherent challenges. To combat such challenges, the NSF requires the panelists to assemble in a two-day face-to-face meeting wherein each proposal is summarized by those who reviewed it and then discussed by the entire panel. This discussion process is essential to the review process yet it is very time-consuming. Below we suggest one way to make these two-day meetings more efficient. The idea is to fairly aggregate the ratings so that one overall global ranked list can be produced. With such an aggregated list, less discussion time can be spent on the bottom, say, one-half of the proposals for which the panel collectively agrees should not be funded.

Let's consider an example. Suppose there are six proposals labeled A through F that must be ranked and four reviewers who are each assigned four proposals to read. A typical panel session begins with the list of Table 14.8. This list nicely compiles reviewer ratings, yet

Table 14.8 Proposal ratings for a sample NSF panel

Proposal	Rating
A	Excellent, Very Good, Excellent
B	Fair, Good
C	Good, Very Good, Very Good
D	Good, Good, Good
E	Good, Excellent
F	Poor, Very Good, Excellent

it hides some important information concerning which reviewers gave which ratings. In fact, the list in Table 14.8 was actually produced from the information in Table 14.9, which clarifies the fact that reviewers use different scales to grade proposals. One reviewer (e.g., in this case,

Table 14.9 Proposal ratings from each reviewer

Reviewer 1		Reviewer 2		Reviewer 3		Reviewer 4	
Proposal	Rating	Proposal	Rating	Proposal	Rating	Proposal	Rating
A	Excellent	A	Very Good	C	Very Good	B	Good
B	Fair	D	Good	E	Excellent	F	Excellent
C	Good	E	Good	F	Very Good	D	Good
D	Good	F	Poor	A	Excellent	C	Very Good

reviewer 2) may be very generous and give a high score of Very Good or Excellent liberally, whereas another may be much stingier and rarely if ever give out an Excellent. As a result, we use the ideas discussed in the section on rating aggregation on 176 to put the ratings of different reviewers on the same scale. Applying the rating-aggregation method produces the following $\bar{\mathbf{R}}$ matrices for the four reviewers.

$$
\bar{\mathbf{R}}_1 = \begin{array}{c} \\ A \\ B \\ C \\ D \\ E \\ F \end{array}\begin{pmatrix} A & B & C & D & E & F \\ 0 & 3/9 & 2/9 & 2/9 & 0 & 0 \\ 0 & 0 & 0 & 0 & 0 & 0 \\ 0 & 1/9 & 0 & 0 & 0 & 0 \\ 0 & 1/9 & 0 & 0 & 0 & 0 \\ 0 & 0 & 0 & 0 & 0 & 0 \\ 0 & 0 & 0 & 0 & 0 & 0 \end{pmatrix}, \quad
\bar{\mathbf{R}}_2 = \begin{array}{c} \\ A \\ B \\ C \\ D \\ E \\ F \end{array}\begin{pmatrix} A & B & C & D & E & F \\ 0 & 0 & 0 & 1/9 & 1/9 & 3/9 \\ 0 & 0 & 0 & 0 & 0 & 0 \\ 0 & 0 & 0 & 0 & 0 & 0 \\ 0 & 0 & 0 & 0 & 0 & 2/9 \\ 0 & 0 & 0 & 0 & 0 & 2/9 \\ 0 & 0 & 0 & 0 & 0 & 0 \end{pmatrix}
$$

$$
\bar{\mathbf{R}}_3 = \begin{array}{c} \\ A \\ B \\ C \\ D \\ E \\ F \end{array}\begin{pmatrix} A & B & C & D & E & F \\ 0 & 0 & 1/4 & 0 & 0 & 1/4 \\ 0 & 0 & 0 & 0 & 0 & 0 \\ 0 & 0 & 0 & 0 & 0 & 0 \\ 0 & 0 & 0 & 0 & 0 & 0 \\ 0 & 0 & 1/4 & 0 & 0 & 1/4 \\ 0 & 0 & 0 & 0 & 0 & 0 \end{pmatrix}, \quad
\bar{\mathbf{R}}_4 = \begin{array}{c} \\ A \\ B \\ C \\ D \\ E \\ F \end{array}\begin{pmatrix} A & B & C & D & E & F \\ 0 & 0 & 0 & 0 & 0 & 0 \\ 0 & 0 & 0 & 0 & 0 & 0 \\ 0 & 1/7 & 0 & 1/7 & 0 & 0 \\ 0 & 0 & 0 & 0 & 0 & 0 \\ 0 & 0 & 0 & 0 & 0 & 0 \\ 0 & 2/7 & 1/7 & 2/7 & 0 & 0 \end{pmatrix}.
$$

The numerator of the (A,B)-entry of $\bar{\mathbf{R}}_1$ means that reviewer 1 rated proposal A three positions above proposal B. The denominator, 9, is the cumulative rating differentials of all $\binom{4}{2} = 6$ possible matchups between proposals rated by reviewer 1. Averaging $\bar{\mathbf{R}}_1, \bar{\mathbf{R}}_2, \bar{\mathbf{R}}_3$, and $\bar{\mathbf{R}}_4$

creates the average distance matrix

$$
\bar{\mathbf{R}}_{ave} = \begin{array}{c} \\ A \\ B \\ C \\ D \\ E \\ F \end{array}
\begin{array}{c}
\begin{array}{cccccc} A & B & C & D & E & F \end{array} \\
\left(\begin{array}{cccccc}
0 & 0.0833 & 0.1181 & 0.0833 & 0.0278 & 0.1458 \\
0 & 0 & 0 & 0 & 0 & 0 \\
0 & 0.0635 & 0 & 0.0357 & 0 & 0 \\
0 & 0.0278 & 0 & 0 & 0 & 0.0556 \\
0 & 0 & 0.0625 & 0 & 0 & 0.1181 \\
0 & 0.0714 & 0.0357 & 0.0714 & 0 & 0
\end{array}\right)
\end{array}.
$$

We used the third method from the section on rating aggregation on 178 to create a rating vector from $\bar{\mathbf{R}}_{ave}$. That is, we computed the Perron vector of $\bar{\mathbf{R}}_{ave}$ and arrived at the rating and ranking vectors appearing in Table 14.10. With these results, the panel might decide to

Table 14.10 Rating-aggregation method applied to NSF proposal example

Proposal	Perron r	Rank
A	.88	1st
B	0	6th
C	0.08	5th
D	0.15	4th
E	0.40	2nd
F	0.20	3rd

discuss only A, E, and F as these are the top three according to the rating-aggregation method.

Summary of Aggregation Methods

- The examples in this chapter demonstrated the use of aggregation methods by combining several *computer-generated* lists. However, these aggregation methods could also be used to combine *human-generated* lists or even *merge* data from both human and computer sources. Consequently, we see great potential for aggregation methods.

- While we have yet to prove a theorem with the following statement, our experiments have led us to posit that $Q_{worst} \leq Q_{aggregate} \leq Q_{best}$, where Q is a quality measure that can be used to score each method. (Part of the problem with proving such a statement lies in the design of an appropriate quality measure Q, which is a nontrivial matter. See Chapter 16.) The statement claims that an aggregated list may be outperformed by the lists produced by one or more individual methods. On the other hand, the aggregated list outperforms the lists produced by several other individual methods. Thus *the aggregated list is helpful in one-time predictive situations*.

ASIDE: March Madness Advice

Consider the annual March Madness tournament and its Bracket Challenge. Predicting a single event is tough and can be hit or miss. The individual (unaggregated) methods in this book predict well "on average," i.e., over the course of many matchups between the same teams. Yet the Bracket Challenge requires that methods predict well in many single occurrence events. This "one-time" prediction problem is different from the "on average" or "best of" prediction problem. In particular, in "one-time" events, upsets are often more frequent due to the heightened, high-stakes circumstances associated with "one and done" style tournaments such as March Madness. Consequently, rather than banking on one particular method, if you must submit one and only one bracket, a wise (and more conservative) choice is to bet on an aggregate of several methods, which is precisely where the rank-aggregation methods of this chapter excel.

By The Numbers —

$47,000,000,000 \approx$ number of Web pages Google has rated and ranked

$12,000,000,000 \approx$ number of Web pages Yahoo! has rated and ranked

$11,000,000,000 \approx$ number of Web pages Bing has rated and ranked

— as of September 2011.

— worldwidewebsize.com

Chapter Fifteen

Rank Aggregation–Part 2

The rank-aggregation methods of the previous chapter are *heuristic* methods, meaning that they come with no guarantees that the aggregated ranking is optimal. On the other hand, the great advantage of these heuristic methods is that they are fast, very fast compared to the optimization method of rank aggregation described in this chapter. Of course, the extra time required by this optimization method is often justified when accuracy is essential.

We now describe one optimal rank-aggregation method, which was created by Dr. Yoshitsugu Yamamoto of the University of Tsukuba in Japan [50]. This method produces an aggregated ranking that optimizes the agreement or conformity among the input rankings. There are several ways to define the agreement between the lists but one possibility creates the constants

$$c_{ij} = (\text{\# of lists with } i \text{ above } j) \quad - \quad (\text{\# of lists with } i \text{ below } j). \tag{15.1}$$

If there are n items in total among the k input lists, then $\mathbf{C}$ is an $n \times n$ skew-symmetric matrix of these constants. $\mathbf{C}$ matrix can be formed to handle input lists that contain ties in their rankings. In addition, $\mathbf{C}$ can be formed for input lists that are either full or partial.

Armed with a matrix $\mathbf{C}$ of constants that measures conformity, our goal is to create a ranking of the n items that maximizes this. In order to accomplish this goal, we define decision variables x_{ij} that determine if item i should be ranked above item j. In particular,

$$x_{ij} = \begin{cases} 1, & \text{if item } i \text{ is ranked above item } j \\ 0, & \text{otherwise.} \end{cases}$$

To understand the use of the matrix $\mathbf{X}$, consider a small example with $n = 4$ items labeled 1 through 4 and ranked in that order. Then the matrix $\mathbf{X}$ associated with this ranking is

$$\mathbf{X} = \begin{matrix} & \begin{matrix} 1 & 2 & 3 & 4 \end{matrix} \\ \begin{matrix} 1 \\ 2 \\ 3 \\ 4 \end{matrix} & \begin{pmatrix} 0 & 1 & 1 & 1 \\ 0 & 0 & 1 & 1 \\ 0 & 0 & 0 & 1 \\ 0 & 0 & 0 & 0 \end{pmatrix} \end{matrix},$$

which indicates that item 1 is ranked above items 2, 3, and 4, while item 2 is ranked above items 3 and 4, and finally, item 3 is ranked above only item 4. In this example, the nice stair-step structure of $\mathbf{X}$ clearly reveals the ranking. At first glance, for other examples, it may not be as clear. Consider the ranking of items from first place to last place given by

[4, 2, 1, 3] with the corresponding matrix

$$
\mathbf{X} = \begin{array}{c} \\ 1 \\ 2 \\ 3 \\ 4 \end{array} \begin{array}{cccc} 1 & 2 & 3 & 4 \\ \left(\begin{array}{cccc} 0 & 0 & 1 & 0 \\ 1 & 0 & 1 & 0 \\ 0 & 0 & 0 & 0 \\ 1 & 1 & 1 & 0 \end{array}\right), \end{array}
$$

which is simply a reordering of the nice stair-step form. That is, reordering $\mathbf{X}$ according to the rank ordering of [4, 2, 1, 3] produces

$$
\text{reordered } \mathbf{X} = \begin{array}{c} \\ 4 \\ 2 \\ 1 \\ 3 \end{array} \begin{array}{cccc} 4 & 2 & 1 & 3 \\ \left(\begin{array}{cccc} 0 & 1 & 1 & 1 \\ 0 & 0 & 1 & 1 \\ 0 & 0 & 0 & 1 \\ 0 & 0 & 0 & 0 \end{array}\right). \end{array}
$$

Fortunately, there is no need to actually reorder the matrix $\mathbf{X}$ resulting from the optimization because it is very easy to identify the ranking from the unordered $\mathbf{X}$. Simply take the column sums of $\mathbf{X}$ and sort them in ascending order to obtain the ranking.[1]

With $\mathbf{X}$ well understood, we now return to the optimization problem. We want to maximize conformity among the input lists, which, in terms of our constants c_{ij} and variables x_{ij} becomes

$$
\max \sum_{i=1}^{n} \sum_{j=1}^{n} c_{ij}\, x_{ij} \quad \text{with } x_{ij} \in \{0, 1\}.
$$

However, we must add some constraints that force the matrix $\mathbf{X}$ to have the properties that we observed and exploited in the small 4×4 example above. This can be accomplished by adding constraints of two types:

$$
x_{ij} + x_{ji} = 1 \quad \text{for all distinct pairs } (i, j) \qquad \text{(Type 1—antisymmetry)}
$$
$$
x_{ij} + x_{jk} + x_{ki} \leq 2 \quad \text{for all distinct triples } (i, j, k) \qquad \text{(Type 2—transitivity).}
$$

The first constraint is an anti-symmetry constraint, which says that exactly one of x_{ij} and x_{ji} can be turned "on" (i.e., set equal to 1). This captures the fact that there are only two choices describing the relationship of i and j: either i is ranked above j or j is ranked above i. The second constraint is a very clever and compact way to enforce transitivity. That is, if $x_{ij} = 1$ (i is ranked above j) and $x_{jk} = 1$ (j is ranked above k), then $x_{ik} = 1$ (i is ranked above k). Transitivity is enforced by the combination of the Type 1 and Type 2 constraints. Because the decision variables are binary, the Type 2 constraint forbids cycles of length 3 from item i back to item i. The Type 1 constraint forbids cycles of length 2. In fact, these two constraints combine to forbid cycles of any length. A *dominance graph* helps explain this.

The matrix $\mathbf{X}$ from our 4×4 example,

$$
\mathbf{X} = \begin{array}{c} \\ 1 \\ 2 \\ 3 \\ 4 \end{array} \begin{array}{cccc} 1 & 2 & 3 & 4 \\ \left(\begin{array}{cccc} 0 & 1 & 1 & 1 \\ 0 & 0 & 1 & 1 \\ 0 & 0 & 0 & 1 \\ 0 & 0 & 0 & 0 \end{array}\right), \end{array}
$$

---

[1] Or the row sums sorted in descending order could be used.

can be alternatively described with the dominance graph of Figure 15.1. Every ranking

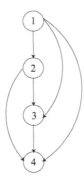

Figure 15.1 Dominance graph

vector produces a graph of this sort, which shows the dominance of an item over all items below it. The dominance graph for every ranking vector will contain no *upward arcs* as an upward arc corresponds to a cycle, i.e., a violation of Type 2 transitivity constraints. To see how the Type 1 and 2 constraints combine to forbid any cycles, consider the cycle from $1 \rightarrow 3 \rightarrow 4 \rightarrow 1$, which corresponds to the Type 2 constraint $x_{13} + x_{34} + x_{41} \leq 2$. Because item 1 is ranked above item 3, x_{13} is turned on (i.e., $x_{13} = 1$). Similarly, $x_{34} = 1$. Then according to the Type 2 constraint, x_{41} must equal 0. Combining this with the Type 1 constraint, then x_{14} must equal 1, and thus, transitivity is enforced. In summary, all three types of constraints (Type 1 and Type 2 plus the binary constraint on x_{ij}) combine to produce an **X** matrix solution that is a simple reordering away from the stair-step form. Finally, because **X** is always a reordering of the stair-step matrix, it has unique row and column sums, and thus, produces a unique ranking of the n items. The complete binary integer linear program (BILP) is

$$\max \sum_{i=1}^{n} \sum_{j=1}^{n} c_{ij}\, x_{ij}$$

$$x_{ij} + x_{ji} = 1 \quad \text{for all distinct pairs } (i,j) \qquad \text{(Type 1–antisymmetry)}$$

$$x_{ij} + x_{jk} + x_{ki} \leq 2 \quad \text{for all distinct triples } (i,j,k) \qquad \text{(Type 2–transitivity)}$$

$$x_{ij} \in \{0,1\} \qquad \text{(Type 3–binary).}$$

Our BILP contains $n(n-1)$ binary decision variables, $n(n-1)$ Type 1 equality constraints, and $n(n-1)(n-2)$ Type 2 inequality constraints. The $O(n^3)$ Type 2 constraints dramatically limit the size of ranking problems that can be solved with this optimal rank-aggregation method. Fortunately, there are some strategies for sidestepping this issue of scale—see the respective sections on constraint relaxation and bounding on pages 190 and 191.

The Running Example

In this chapter we deviate from our usual 5×5 running example and introduce another slightly bigger 12×12 example. This example, which comes from the 2008–2009 Southern Conference (SoCon) basketball season, contains some interesting properties that the 5×5

example does not. In particular the SoCon example is more appropriate because it contains ties, which is the subject of the section on multiple optimal solutions on page 187.

In this section, we aggregate the rankings created by 16 different methods. For the input methods, we used the Colley, Massey, Markov, and OD methods, each with various weightings. Using the definition for conformity from Equation (15.1), the $\mathbf{C}$ matrix is

$$
\mathbf{C} =
\begin{array}{c}
 \\
1 \\ 2 \\ 3 \\ 4 \\ 5 \\ 6 \\ 7 \\ 8 \\ 9 \\ 10 \\ 11 \\ 12
\end{array}
\begin{pmatrix}
\begin{array}{cccccccccccc}
1 & 2 & 3 & 4 & 5 & 6 & 7 & 8 & 9 & 10 & 11 & 12 \\
0 & -16 & -14 & -12 & -12 & 16 & 14 & 14 & -2 & 16 & -14 & -14 \\
16 & 0 & -12 & -12 & -12 & 16 & 16 & 16 & 14 & 16 & 12 & 0 \\
14 & 12 & 0 & -12 & -12 & 16 & 16 & 16 & 16 & 16 & 12 & 12 \\
12 & 12 & 12 & 0 & -14 & 14 & 12 & 12 & 12 & 14 & 12 & 12 \\
12 & 12 & 12 & 14 & 0 & 16 & 12 & 12 & 12 & 12 & 12 & 12 \\
-16 & -16 & -16 & -14 & -16 & 0 & 12 & 12 & -16 & 12 & -16 & -16 \\
-14 & -16 & -16 & -12 & -12 & -12 & 0 & -12 & -14 & 6 & -12 & -14 \\
-14 & -16 & -16 & -12 & -12 & -12 & 12 & 0 & -14 & 16 & -12 & -14 \\
2 & -14 & -16 & -12 & -12 & 16 & 14 & 14 & 0 & 16 & -12 & -16 \\
-16 & -16 & -16 & -14 & -12 & -12 & -6 & -16 & -16 & 0 & -16 & -16 \\
14 & -12 & -12 & -12 & -12 & 16 & 12 & 12 & 12 & 16 & 0 & -8 \\
14 & 0 & -12 & -12 & -12 & 16 & 14 & 14 & 16 & 16 & 8 & 0
\end{array}
\end{pmatrix}.
$$

Solving the BILP produces an optimal objective value of 870 with the solution matrix $\mathbf{X}$ below.

$$
\mathbf{X} =
\begin{array}{c}
1 \\ 2 \\ 3 \\ 4 \\ 5 \\ 6 \\ 7 \\ 8 \\ 9 \\ 10 \\ 11 \\ 12
\end{array}
\begin{pmatrix}
\begin{array}{cccccccccccc}
1 & 2 & 3 & 4 & 5 & 6 & 7 & 8 & 9 & 10 & 11 & 12 \\
0 & 0 & 0 & 0 & 0 & 1 & 1 & 1 & 0 & 1 & 0 & 0 \\
1 & 0 & 0 & 0 & 0 & 1 & 1 & 1 & 1 & 1 & 1 & 0 \\
1 & 1 & 0 & 0 & 0 & 1 & 1 & 1 & 1 & 1 & 1 & 1 \\
1 & 1 & 1 & 0 & 0 & 1 & 1 & 1 & 1 & 1 & 1 & 1 \\
1 & 1 & 1 & 1 & 0 & 1 & 1 & 1 & 1 & 1 & 1 & 1 \\
0 & 0 & 0 & 0 & 0 & 0 & 1 & 1 & 0 & 1 & 0 & 0 \\
0 & 0 & 0 & 0 & 0 & 0 & 0 & 0 & 0 & 1 & 0 & 0 \\
0 & 0 & 0 & 0 & 0 & 0 & 1 & 0 & 0 & 1 & 0 & 0 \\
1 & 0 & 0 & 0 & 0 & 1 & 1 & 1 & 0 & 1 & 0 & 0 \\
0 & 0 & 0 & 0 & 0 & 0 & 0 & 0 & 0 & 0 & 0 & 0 \\
1 & 0 & 0 & 0 & 0 & 1 & 1 & 1 & 1 & 1 & 0 & 0 \\
1 & 1 & 0 & 0 & 0 & 1 & 1 & 1 & 1 & 1 & 1 & 0
\end{array}
\end{pmatrix}.
$$

Sorting the column sums of $\mathbf{X}$ in ascending order ranks the twelve SoCon teams as

$$
\begin{pmatrix}
\text{Davidson} & 5 \\
\text{CofC} & 4 \\
\text{Citadel} & 3 \\
\text{Wofford} & 12 \\
\text{UT Chatt} & 2 \\
\text{W. Carolina} & 11 \\
\text{Samford} & 9 \\
\text{App State} & 1 \\
\text{Elon} & 6 \\
\text{GA Southern} & 8 \\
\text{Furman} & 7 \\
\text{UNC-G} & 10
\end{pmatrix}.
$$

Solving the BILP

BILPs are typically solved with a technique called branch and bound, which uses a series of linear programming (LP) relaxations of the problem to form a tree to narrow down the process of stepping through the discrete solution space. Because discrete optimization is much harder than continuous optimization, the size of the problem, becomes an issue. Remember the size of the problem is governed by n, the number of items being ranked. Our rank-aggregation BILP contains $n(n-1)$ binary decision variables as well as $n(n-1)$ Type 1 constraints and $n(n-1)(n-2)$ Type 2 constraints. The $O(n^3)$ number of constraints dramatically limits the size of the ranking problem that can be handled by this optimal conformity method. In contrast, the heuristic method of the previous chapter had no such scalability issues. Fortunately, there are some very clever strategies for sidestepping this scale issue, as the discussion in the sections on pages 190 and 191 show.

Multiple Optimal Solutions for the BILP

The branch and bound procedure terminates with an optimal solution $\mathbf{X}$. As we saw with our small examples, finding and sorting the column sums in ascending order gives the optimal ranking of the items. Problem solved—we can pack up and go home. Not if you are a mathematician. Mathematicians like to ask deeper questions, such as: is the optimal solution unique? If not, can we find alternate optimal solutions?

It turns out that these questions have answers that reveal hints for circumventing the scalability issues mentioned above. There is a simple test to determine if the optimal solution to the BILP is unique. Consider each successive pair of items in the optimal ranked list and ask if the two items i and j can be swapped without changing the objective value. Only swaps of rank neighboring items need be considered as these are the only swaps that do not violate the constraints. The objective value will not change if $c_{ij} = c_{ji}$. If this is so, then an alternate optimal solution is one that has these two items swapped. There may indeed be more than a two-way tie at this rank position. For instance, a three-way tie occurs if $c_{ij} = c_{ji} = c_{ik} = c_{ki} = c_{jk} = c_{kj}$ for rank neighboring items i, j, and k. Continue down the optimal ranked list in this fashion detecting any two-way or higher ties at each position. We apply this Tie Detection algorithm to the SoCon example. From the running example on page 185, the BILP produced the optimal ranking of

$$
\begin{array}{l|c}
\text{Davidson} & 5 \\
\text{CofC} & 4 \\
\text{Citadel} & 3 \\
\text{Wofford} & 12 \\
\text{UT Chatt} & 2 \\
\text{W. Carolina} & 11 \\
\text{Samford} & 9 \\
\text{App State} & 1 \\
\text{Elon} & 6 \\
\text{GA Southern} & 8 \\
\text{Furman} & 7 \\
\text{UNC-G} & 10
\end{array}
$$

We begin the Tie Detection algorithm by comparing the two teams at the top of the list. Because $\mathbf{C}(5,4) \neq \mathbf{C}(4,5)$, these two cannot be swapped. Thus, we move onto the next pair of teams in the ranked list, teams 4 and 3. Because $\mathbf{C}(4,3) \neq \mathbf{C}(3,4)$, these two cannot be swapped either. We move down the list to teams 3 and 12, which again cannot be swapped. Finally, we reach a pair that can be swapped, the pair of 12 and 2. This means that teams 12 and 2 can appear in either order in the optimal ranked list, with 12 above 2 as the BILP algorithm found or with 2 above 12, which is an alternate optimal solution since the objective function is unchanged yet feasibility is still maintained. At this point, we have discovered a two-way tie between teams 12 and 2, but a three-way or higher tie may exist. So we check to see if the next team in the list, team 11, satisfies

$C(12, 2) = C(2, 12) = C(2, 11) = C(11, 2) = C(12, 11) = C(11, 12)$, which it does not. Thus, the tie at the fourth rank position is indeed only a two-way tie between teams 12 and 2. We continue down the list, considering 11 and 9, then 9 and 1, and so on, and we find no more ties. As a result, this SoCon example has two integer optimal solutions and infinitely many fractional optional solutions on the boundary between these two.

In summary, we know that we can (1) apply a branch and bound procedure to find an optimal solution to the rank-aggregation BILP, (2) check the uniqueness of the obtained optimal solution, and (3) if applicable, find *several* alternate optimal solutions with the simple $O(n)$ test described above. As a result, this optimization technique produces an output ranking that may actually contain ties and is a very mathematically appealing and provably optimal ranking method. However, the BILP is much slower than many existing rating and ranking methods. In fact, because of the $O(n^3)$ constraints, in practice, a commercial BILP solver such as the DASH Optimization software or the NEOS server is limited to a problem with about a thousand items. Thus, ranking all NCAA Division 1 football or basketball teams is certainly within reach while ranking billions of webpages in cyberspace is not. Yet ranking the top 50 results produced by several popular search engines is not only possible, but fast—as it can be done in real-time. Fortunately, the next section explains how we can further increase the practical limit on n, making the aggregated ranking of several thousand items possible.

The LP Relaxation of the BILP

The first step to solving a big nasty BILP is to solve a related problem, the relaxed LP. For us, this means relaxing the Type 3 constraints that force the variables x_{ij} to be discrete, specifically, binary since $x_{ij} \in \{0, 1\}$, and allow them to be continuous so that $0 \leq x_{ij} \leq 1$. Actually, the upper bound of the bound constraint $0 \leq x_{ij} \leq 1$ is redundant as this restriction is covered by the Type 1 constraint $x_{ij} + x_{ji} = 1$. Thus, the simplified relaxed LP for the rank-aggregation problem is

$$\max \sum_{i=1}^{n} \sum_{j=1}^{n} c_{ij} \, x_{ij}$$

$$
\begin{aligned}
x_{ij} + x_{ji} &= 1 && \text{for all distinct pairs } (i, j) && \text{(Type 1–antisymmetry)} \\
x_{ij} + x_{jk} + x_{ki} &\leq 2 && \text{for all distinct triples } (i, j, k) && \text{(Type 2–transitivity)} \\
x_{ij} &\geq 0 && && \text{(Type 3–\textit{continuous})}.
\end{aligned}
$$

When some BILPs are solved as LPs the optimal solution to the relaxed problem, the LP, gives a solution with binary values, which is clearly also optimal for the BILP. This is the best-case scenario. The next best scenario is when the optimal solution for the LP contains just a small proportion of fractional values. Often, in this case, these few fractional values can be rounded to the nearest integer giving a slightly suboptimal solution that, if feasible, may adequately approximate the exact optimal integer solution. In this section, we show that the LP gives very interesting results. Many times the LP solution is optimal and, in fact, readily tells us all alternate optimal solutions as well.

Our 12-team SoCon example makes this point well. We discovered in the section on multiple optimal solutions on 187 that this example has two integer optimal solutions. One with the teams in the rank order given by $(\begin{matrix} 5 & 4 & 3 & 12 & 2 & 11 & 9 & 1 & 6 & 8 & 7 & 10 \end{matrix})$

and the other with teams 12 and 2 switched. Notice how this tie is manifested in the LP solution matrix $\mathbf{X}$ below.

$$
\mathbf{X} = \begin{array}{c} \\ 1 \\ 2 \\ 3 \\ 4 \\ 5 \\ 6 \\ 7 \\ 8 \\ 9 \\ 10 \\ 11 \\ 12 \end{array}
\begin{array}{c}
\begin{array}{cccccccccccc} 1 & 2 & 3 & 4 & 5 & 6 & 7 & 8 & 9 & 10 & 11 & 12 \end{array} \\
\left(\begin{array}{cccccccccccc}
0 & 0 & 0 & 0 & 0 & 1 & 1 & 1 & 0 & 1 & 0 & 0 \\
1 & 0 & 0 & 0 & 0 & 1 & 1 & 1 & 1 & 1 & 1 & .478 \\
1 & 1 & 0 & 0 & 0 & 1 & 1 & 1 & 1 & 1 & 1 & 1 \\
1 & 1 & 1 & 0 & 0 & 1 & 1 & 1 & 1 & 1 & 1 & 1 \\
1 & 1 & 1 & 1 & 0 & 1 & 1 & 1 & 1 & 1 & 1 & 1 \\
0 & 0 & 0 & 0 & 0 & 0 & 1 & 1 & 0 & 1 & 0 & 0 \\
0 & 0 & 0 & 0 & 0 & 0 & 0 & 0 & 0 & 1 & 0 & 0 \\
0 & 0 & 0 & 0 & 0 & 0 & 1 & 0 & 0 & 1 & 0 & 0 \\
1 & 0 & 0 & 0 & 0 & 1 & 1 & 1 & 0 & 1 & 0 & 0 \\
0 & 0 & 0 & 0 & 0 & 0 & 0 & 0 & 0 & 0 & 0 & 0 \\
1 & 0 & 0 & 0 & 0 & 1 & 1 & 1 & 1 & 1 & 0 & 0 \\
1 & .522 & 0 & 0 & 0 & 1 & 1 & 1 & 1 & 1 & 1 & 0
\end{array} \right)
\end{array}.
$$

The LP solution contains fractional values in locations $\mathbf{X}(12, 2)$ and $\mathbf{X}(2, 12)$, precisely the locations in $\mathbf{C}$ that indicated a tie between teams 12 and 2. The LP solver has terminated with a fractional optimal solution and this solution lies on the boundary between the two integer optimal solutions of the BILP. Because the LP terminated with the identical optimal objective value, 870, that the BILP has, we know the LP solution is optimal for the original BILP.

Though the above SoCon example ends in a fractional solution from which optimal binary solutions can be constructed, this is not guaranteed. In fact, we constructed a 9-item example with a unique fractional optimal solution. Thus, when binary solutions are constructed, the objective value is not as good as that produced by the unique fractional solution. However, this example with a unique fractional solution was hard to construct. In fact, to locate such an instance, we randomly generated thousands of $\mathbf{C}$ matrices with entries uniformly distributed between -1 and 1 before we encountered the unique fractional 9-item example. Thus, though unique fractional solutions are possible, it is more likely in practice that the relaxed LP formulation of our ranking problem will result in non-unique fractional solutions from which multiple optimal binary solutions can be constructed. Empirical studies by Reinelt et al. [62, 65, 64, 63] add further evidence that the LP results for ranking problems are exceptionally good and often optimal and binary in practice.

In which cases can we be certain that the optimal LP solution is also optimal for the BILP? Remember, after all, that the BILP is truly the problem of interest for us. Of course, if the LP solution is binary, then that solution is optimal for the BILP. But even when the LP solution is fractional, there are instances in which it is optimal for the BILP as the SoCon example showed. In [50], Langville, Pedings, and Yamamoto proved conditions that identify when an LP solution is optimal for the BILP. Their theorem also connects the presence of multiple optimal solutions for the BILP, which indicate the presence of ties in the ranking, to fractional values in the LP solution.

Their theorem carries computational consequences as well. The original LP solver, the famous simplex method, is not the method of choice for solving our MVR problem. The simplex method is an extreme point method, meaning that it moves from one extreme point to the next, in an ever-improving direction until it reaches an optimal solution, which will, of course, also be an extreme point. In contrast, interior point LP solvers move through the interior of the feasible region until they converge on an optimal solution that may be an extreme point or a boundary point depending on the path taken through the interior of the feasible region. For us, it is the boundary optimal points (which contain fractional values) that give us so much more information than extreme optimal points (which are the integer-only solutions). Thus, we always choose a non-extreme point, non-simplex LP solver when solving the relaxed LP associated with the ranking problem.

Constraint Relaxation

Even with the LP relaxation, the $O(n^3)$ Type 2 constraints drastically limit the size of the ranking problems that we can handle. So we must resort to another relaxation technique, *constraint relaxation*, to increase the size of tractable problems. Of the $n(n-1)(n-2)$ Type 2 constraints, $x_{ij}+x_{jk}+x_{ki} \leq 2$, only a very small proportion of these are necessary. The great majority of these will be trivially satisfied—the problem is that we don't know which are necessary and which are unnecessary. In order to find out, we start by assuming all are unnecessary, then slowly add back in the necessary ones as they are identified. The Type 1 and relaxed continuous version of the Type 3 constraints cause no problems, so we leave these unchanged. Below are the steps involved in the constraint relaxation technique. Constraint relaxation can be applied with equal success to either the BILP or the LP. We have chosen to describe it for the BILP only.

CONSTRAINT RELAXATION ALGORITHM FOR LARGE RANKING BILPS

1. Relax all Type 2 constraints so that the initial set of necessary Type 2 constraints is empty.

2. Solve the BILP with the current set of necessary Type 2 constraints. Form the optimal ranking associated with this solution. This ranking is actually an approximation to the true ranking we desire for the original problem with the full Type 2 constraint set.

3. Determine which Type 2 constraints are violated by the solution from Step 2—these are *necessary* Type 2 constraints. Add these Type 2 constraints to the set of necessary Type 2 constraints and go to Step 2. Repeat until no Type 2 constraints are violated. The BILP solution at this point is an optimal ranking for the original problem with the full constraint set.

In Step 3, the determination of which Type 2 constraints are violated by the current BILP solution is easy and does not require that each constraint be checked one by one. Recall from Figure 15.1 that violations to the Type 2 transitivity constraints are *upward arcs* in the *dominance graph*. In yet another view, these violations manifest as ones on the lower triangular part of the rank reordered matrix $\mathbf{X}$. For each identified upward arc (j, i), next find all k such that $x_{ik} = x_{kj} = 1$ and generate the corresponding Type 2 constraint $x_{ji} + x_{ik} + x_{kj} \leq 2$. The matrix $\mathbf{X}$ can be used to quickly find these elements k: form the Hadamard (element-wise) product of the ith row and j^{th} column of $\mathbf{X}$. All nonzero elements in this product satisfy $x_{ik} = x_{kj} = 1$. Because of the upper triangular structure of the reordered matrix $\mathbf{X}$, it takes much less than $O(n)$ work to compute the Hadamard product and form the transitivity constraints associated with each upward arc.

In addition, we can take advantage of any approximate rankings that may exist. For example, suppose a fast heuristic method is run and a ranking produced. This ranking has a one-to-one correspondence to a matrix, let's call it $\bar{\mathbf{X}}$, in upper triangular form. Computing the objective value $f(\bar{\mathbf{X}}) = \mathbf{C} \cdot * \bar{\mathbf{X}}$ for this approximate solution matrix $\bar{\mathbf{X}}$ gives a useful lower bound on the objective. As the branch and bound BILP procedure explores solutions and encounters a branch with nodes having value below $f(\bar{\mathbf{X}})$, nodes in that branch no longer need to be explored.

In summary, the constraint relaxation technique is an iterative procedure that solves a series of BILPs whose constraint set gradually grows until all the necessary transitivity constraints are identified. At each iteration the optimal BILP solution is an approximation to the true optimal ranking of the original problem with the full constraint set. The approximations improve until the optimal ranking is reached.

Sensitivity Analysis

Another advantage of the LP over the BILP relates to the natural sensitivity measures produced when solving a linear program. In this case, we are interested in changes in the objective coefficients c_{ij}. Slight changes in the input data (specifically the differential matrix $\mathbf{D}$ that creates the objective coefficient matrix $\mathbf{C}$) could change the optimal solution, and hence, optimal ranking of the teams. Industrial optimization software, such as Xpress-MP, computes the following ranges given in Figure 15.2 on the objective coefficients for the 12-team SoCon example.

```
        1        2         3        4        5        6         7        8        9        10        11        12
-------------------------------------------------------------------------------------------------------------------
 1 : [   0,   0][-Inf,  16][-Inf,  14][-Inf,  12][-Inf,  12][ -16, Inf][ -14, Inf][ -14, Inf][-Inf,   2][ -16, Inf][-Inf,  14][-Inf,  14]
 2 : [ -16, Inf][   0,   0][-Inf,  12][-Inf,  12][-Inf,  12][ -16, Inf][ -16, Inf][ -16, Inf][ -14, Inf][ -16, Inf][-Inf, -12, Inf][  -0,  32]
 3 : [ -14, Inf][ -12, Inf][   0,   0][-Inf,  12][-Inf,  12][ -16, Inf][ -16, Inf][ -16, Inf][ -16, Inf][ -16, Inf][-Inf, -12, Inf][ -12, Inf]
 4 : [ -12, Inf][ -12, Inf][ -12, Inf][   0,   0][-Inf,  14][ -14, Inf][ -12, Inf][ -12, Inf][ -12, Inf][ -14, Inf][-Inf, -12, Inf][ -12, Inf]
 5 : [ -12, Inf][ -12, Inf][ -12, Inf][ -14, Inf][   0,   0][ -16, Inf][ -12, Inf][ -12, Inf][ -12, Inf][ -12, Inf][-Inf, -12, Inf][ -12, Inf]
 6 : [-Inf,  16][-Inf,  16][-Inf,  16][-Inf,  14][-Inf,  16][   0,   0][ -12, Inf][ -12, Inf][-Inf,  16][ -12, Inf][-Inf,  16][-Inf,  16]
 7 : [-Inf,  14][-Inf,  16][-Inf,  16][-Inf,  12][-Inf,  12][-Inf,  12][   0,   0][-Inf,  12][-Inf,  14][  -6, Inf][-Inf,  12][-Inf,  14]
 8 : [-Inf,  14][-Inf,  16][-Inf,  16][-Inf,  12][-Inf,  12][ -12, Inf][   0,   0][-Inf,  14][ -16, Inf][-Inf,  12][-Inf,  14]
 9 : [  -2, Inf][-Inf,  14][-Inf,  16][-Inf,  12][-Inf,  12][ -16, Inf][ -14, Inf][ -14, Inf][   0,   0][ -16, Inf][-Inf,  12][-Inf,  16]
10 : [-Inf,  16][-Inf,  16][-Inf,  14][-Inf,  14][-Inf,  12][ -12, Inf][   6][-Inf,  16][   0,   0][-Inf,  16][-Inf,  16]
11 : [ -14, Inf][-Inf,  12][-Inf,  12][-Inf,  12][-Inf,  12][ -16, Inf][ -12, Inf][ -12, Inf][ -16, Inf][   0,   0][-Inf,   8]
12 : [ -14, Inf][ -32,   0][-Inf,  12][-Inf,  12][-Inf,  12][ -16, Inf][ -14, Inf][ -14, Inf][ -16, Inf][ -16, Inf][  -8, Inf][   0,   0]
```

Figure 15.2 Sensitivity ranges on the objective coefficients c_{ij}

Most c_{ij} coefficients have very loose bounds. In fact, the only entries with finite ranges are the pair of 2 and 12. For this dataset, these ranges warn us that we are less certain of the rank ordering of these two teams in the middle of the pack, which are the same two teams that, in the section on multiple optimal solutions on 187, we found are tied in the ranking. Changes of the objective coefficients outside of the given ranges, can change the ranking.

Bounding

Bounding techniques are extremely useful in optimization. In particular, solution techniques for integer programs rely heavily on bounding. In this section, we apply such bounding to the Iterative LP method to accelerate convergence and produce optimality guarantees. Recall that the Iterative LP method relaxes the transitivity constraint set. In fact, at the first iteration, the LP is solved with no transitivity constraints. The optimal solution matrix $\mathbf{X}$ at this iteration creates an objective function value that we denote $\bar{f}$

since it is an upper bound of the optimal objective value f^*. The solution at this iteration almost always violates some transitivity constraints and thus is not a feasible solution for the original LP. However, it can be used to form a very useful approximate solution. This is done by computing the row sums of X. The ith row sum is a good indicator of how many opponents the ith team will beat. As the iterative LP method proceeds, we use the row sums of each iteration's solution matrix X to compute an approximate ranking. This approximation gets closer to the optimal ranking as the iterations proceed. Because every ranking, including these approximate ones, has a one-to-one correspondence with an X matrix, we can compute the objective function value for each approximate ranking, which we denote $\underline{f}$ since it is a lower bound of f^*. Thus, we know that

$$\underline{f} \le f^* \le \overline{f}.$$

Next we bound the relative error $\frac{\overline{f} - f^*}{f^*}$ associated with the approximate ranking. Because $\underline{f}$, f^*, and $\overline{f}$ all have the same sign, we can bound the relative error, which involves the unknown f^* with the known quantities $\underline{f}$ and $\overline{f}$.

$$\frac{\overline{f} - f^*}{|f^*|} \le \frac{\overline{f} - \underline{f}}{|f^*|} \le \frac{\overline{f} - \underline{f}}{|\overline{f}|}.$$

Further, when all elements of the objective coefficient matrix C are integral and $\overline{f} - \underline{f} \le 1$, the approximate ranking is the optimal solution of the original ranking problem. Even when we are not so lucky as to be able to guarantee optimality, we can give a guarantee on the error associated with the near-optimal solution. We can guarantee that the optimal objective value is between $\underline{f}$ and $\overline{f}$ and the relative error is not greater than $\frac{\overline{f} - \underline{f}}{|\overline{f}|}$. This is another great advantage of the bounding version of our Iterative LP method. Not only does it require fewer iterations but it also allows the user to stop the iterative procedure as soon as an acceptable relative error is reached.

ASIDE: Ranking in Rome

Throughout this book we have emphasized the many and varied uses of the ranking methods described. Perhaps one of the most unlikely and little known uses is archaeology. Picture the large excavation sites of Pompeii or Rome, Italy. The phrase "when in Rome, rank" is good advice for an archaeologist hoping to correctly date items from a dig. A typical dig is divided into several sites at which excavation is done. At each site, recovered items are carefully tagged according to the depth from which they were unearthed. Multiple styles of pottery or classes of artifacts are discovered at multiple sites at multiple dig depths. The archaeologist's goal is to determine the relative age of a class of artifact. For instance, for transporting water, did the Romans first use painted clay vases or mosaic mud jugs? At one dig site the vase may have appeared at a lower dig depth than the jug, suggesting the vase came first. The order may be reversed at another site. In effect, each site creates a ranking of the items according to their dig depth. The goal is to determine the overall ranking of the artifacts that is most consistent with all sites. Mathematically, we want to aggregate the rankings from each site into an optimal ranking that maximizes agreement among the many sites.

Summary of the Rank-Aggregation (by Optimization) Method

The shaded box below assembles the labeling conventions we have adopted to describe our optimal rank-aggregation method.

Notation for the Rank-Aggregation (by Optimization) Method

n number of items to be ranked

$\mathbf{C}$ conformity matrix with coefficients of objective function

---

The list below summarizes the steps of the rank-aggregation method.

RANK-AGGREGATION (BY OPTIMIZATION) METHOD

1. Form the matrix $\mathbf{C}$ according to some conformity definition. For example,

$$c_{ij} = (\text{\# of lists with } i \text{ above } j) \;-\; (\text{\# of lists with } i \text{ below } j).$$

2. Solve the LP optimization problem

$$\max \sum_{i=1}^{n} \sum_{j=1}^{n} c_{ij}\, x_{ij}$$

$$x_{ij} + x_{ji} = 1 \quad \text{for all distinct pairs } (i,j) \qquad \text{(Type 1–antisymmetry)}$$
$$x_{ij} + x_{jk} + x_{ki} \leq 2 \quad \text{for all distinct triples } (i,j,k) \qquad \text{(Type 2–transitivity)}$$
$$x_{ij} \geq 0 \qquad \text{(Type 3–\textit{continuous})}.$$

For problems with large n, use the constraint relaxation and bounding techniques discussed in the sections on pages 190 and 191.

3. The optimal ranking is produced by sorting the columns of the optimal solution matrix $\mathbf{X}$ in ascending order. Ties in the ranking are identified by the location of fractional values in $\mathbf{X}$.

We close this section with a list of properties of this rank-aggregation method.

- Our experiments show that when the lists to be aggregated are largely in agreement from the start, then only a handful of iterations are required. This is very good news if you are in the business of building a meta-search engine that aggregates the top-100 results from five popular search engines. Our rank-aggregation methods can be executed in just a few seconds on the latest laptop machine.

- The constraints of the simplified relaxed LP create a polytope that has been well-studied and is often referred to as the linear ordering polytope [65]. It is helpful to study the relationship of the LP's polytope to the feasible region of the original BILP. Of course, the feasible region of the BILP is contained within the feasible region of the LP. The best scenario is when the LP's feasible region is as tight as possible to the BILP's feasible region. In other words, the LP's feasible region is the convex hull of the points in the BILP's feasible region. For our ranking problem, the good news is that all of the inequality constraints (Type 2 transitivity and Type 3 nonnegativity) are facet-defining inequalities for the linear ordering polytope. This means that *these* inequalities are as tight as possible. However, the set of constraints for the LP does not cover all facet-defining inequalities for the linear ordering polytope. Sophisticated valid inequalities such as the so-called fence and Mobius ladder inequalities create stronger LP relaxations, but unfortunately they are too costly to generate [57, 62, 64].

- LPs come with the wonderful capability of sensitivity analysis. That is, LPs are amenable to "what if" scenario analysis. When certain input parameters are changed, we can determine if the optimal solution changes and by how much. Because we have shown that an LP can be used to solve the ranking problem, then there is the potential for sensitivity analysis. In this case, we can answer questions about how changes in the objective coefficients c_{ij} affect the optimal solution. Such sensitivity analysis is additional information that few, if any, other ranking methods deliver.

Revisiting the Rating-Differential Method

In Chapter 8, we described the rating-differential method for ranking items. That method aims to find a reordering of the items that when applied to the item-item matrix of differential data forms a matrix that is as close to the so-called *hillside form* as possible. The following figure summarizes the concept pictorially.

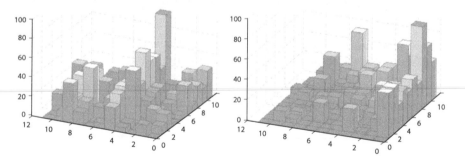

On the left is a cityplot of an 11×11 matrix in its original ordering of items, while the right is a cityplot of the same data displayed with the new optimal ordering. The cityplot on the right resembles a hillside as the ordering was chosen to bring the matrix as close to hillside form as possible.

Rating Differential vs. Rank Aggregation

In Chapter 8 we formulated an optimization problem whose objective was to minimize the (possibly weighted) count of the number of violations from the hillside form defined in the box on page 108. In that chapter, we described an evolutionary approach to solving this optimization problem. In this section, we show that the rating-differential optimization problem can actually be solved optimally and efficiently with the techniques of this chapter. Much of the work for this section was developed in [50], where we referred to the method as the Minimum Violation Ranking (MVR) method, since it produces a ranking that minimizes hillside violations.

In fact, the rating-differential method fits exactly into the BILP, and thus, LP, formulation of this chapter. Let's focus on the data matrix $\mathbf{D}$ of point differentials, wherein d_{ij} is the number of points winning team i beat losing team j by in their matchup, 0 otherwise. The trick is to think of each row and column of $\mathbf{D}$ as a team's ranking of its opponents. For instance, for our 5-team example with

$$
\mathbf{D} = \begin{array}{c} \\ \text{Duke} \\ \text{Miami} \\ \text{UNC} \\ \text{UVA} \\ \text{VT} \end{array}
\begin{array}{ccccc}
\text{Duke} & \text{Miami} & \text{UNC} & \text{UVA} & \text{VT} \\
\left(\begin{array}{ccccc}
0 & 0 & 0 & 0 & 0 \\
45 & 0 & 18 & 8 & 20 \\
3 & 0 & 0 & 2 & 0 \\
31 & 0 & 0 & 0 & 0 \\
45 & 0 & 27 & 38 & 0
\end{array} \right)
\end{array},
$$

the second row of $\mathbf{D}$ tells us that Miami would rank its opponents' defensive ability from strongest to weakest as UVA, UNC, VT, Duke. On the other hand, the first column of $\mathbf{D}$, for example, tells us that Duke would rank its opponents' offensive ability as Miami/VT, UVA, UNC. Consequently, these rankings, both offensive and defensive for the n teams can be aggregated to create an overall ranking for the season. The only modification to the rank-aggregation method of this chapter is in the definition of the conformity matrix $\mathbf{C}$. Rather than the definition of Equation (15.1), the conformity matrix for the rating-differential method is

---

Definition of the C Matrix

Let $\mathbf{C} = [c_{ij}] \ \forall \ i, j = 1, 2, \ldots, n$ be defined as

$$c_{ij} := \#\{ k \mid d_{ik} < d_{jk} \} + \#\{ k \mid d_{ki} > d_{kj} \}, \tag{15.2}$$

where $\#$ denotes the cardinality of the corresponding set. Thus,

$$\#\{ k \mid d_{ik} < d_{jk} \}$$

is the number of teams receiving a lower point differential against team i than team j. Similarly, $\#\{ k \mid d_{ki} > d_{kj} \}$ is the number of teams receiving a greater point differential against team i than team j.

Note: The matrix $\mathbf{C}$ above counts hillside violations in a binary fashion, however, something more sophisticated can be done. For instance, we can consider weighted violations by summing the difference each time a hillside violation occurs. In this case, $\mathbf{C}$ is defined as $c_{ij} := \sum_{k:d_{ik}<d_{jk}} (d_{jk} - d_{ik}) + \sum_{k:d_{ki}>d_{kj}} (d_{ki} - d_{kj})$.

The Running Example

This definition for C produces the following conformity matrix for the 12-team SoCon example.

$$
C = \begin{array}{c} \\ 1 \\ 2 \\ 3 \\ 4 \\ 5 \\ 6 \\ 7 \\ 8 \\ 9 \\ 10 \\ 11 \\ 12 \end{array}
\begin{array}{c}
\begin{array}{cccccccccccc}
1 & 2 & 3 & 4 & 5 & 6 & 7 & 8 & 9 & 10 & 11 & 12
\end{array} \\
\left(\begin{array}{cccccccccccc}
0 & 15 & 15 & 14 & 17 & 7 & 4 & 4 & 9 & 2 & 10 & 11 \\
8 & 0 & 10 & 12 & 18 & 6 & 3 & 3 & 11 & 3 & 7 & 8 \\
5 & 11 & 0 & 9 & 14 & 6 & 2 & 4 & 9 & 2 & 5 & 9 \\
5 & 9 & 9 & 0 & 15 & 5 & 0 & 2 & 6 & 3 & 6 & 5 \\
2 & 2 & 5 & 3 & 0 & 2 & 1 & 2 & 0 & 1 & 1 & 2 \\
10 & 14 & 16 & 17 & 18 & 0 & 7 & 7 & 12 & 4 & 13 & 15 \\
15 & 18 & 18 & 20 & 20 & 13 & 0 & 8 & 16 & 10 & 15 & 15 \\
15 & 20 & 18 & 18 & 20 & 13 & 10 & 0 & 15 & 11 & 14 & 18 \\
10 & 9 & 11 & 14 & 19 & 7 & 4 & 7 & 0 & 2 & 10 & 9 \\
17 & 17 & 18 & 18 & 20 & 16 & 7 & 9 & 15 & 0 & 13 & 14 \\
10 & 14 & 14 & 10 & 18 & 8 & 4 & 4 & 12 & 7 & 0 & 12 \\
10 & 12 & 11 & 12 & 17 & 7 & 4 & 4 & 10 & 6 & 8 & 0
\end{array}\right)
\end{array}.
$$

Solving this example with LP formulation of this chapter produces the optimal X matrix below.

$$
X =
\begin{array}{c}
\begin{array}{cccccccccccc}
1 & 2 & 3 & 4 & 5 & 6 & 7 & 8 & 9 & 10 & 11 & 12
\end{array} \\
\left(\begin{array}{cccccccccccc}
0 & 0 & 0 & 0 & 0 & 1 & 1 & 1 & 0 & 1 & 0.4951 & 0 \\
1 & 0 & 0 & 0 & 0 & 1 & 1 & 1 & 0 & 1 & 1 & 1 \\
1 & 1 & 0 & 0.5896 & 0 & 1 & 1 & 1 & 1 & 1 & 1 & 1 \\
1 & 1 & 0.4104 & 0 & 0 & 1 & 1 & 1 & 1 & 1 & 1 & 1 \\
1 & 1 & 1 & 1 & 0 & 1 & 1 & 1 & 1 & 1 & 1 & 1 \\
0 & 0 & 0 & 0 & 0 & 0 & 1 & 1 & 0 & 1 & 0 & 0 \\
0 & 0 & 0 & 0 & 0 & 0 & 0 & 1 & 0 & 0 & 0 & 0 \\
0 & 0 & 0 & 0 & 0 & 0 & 0 & 0 & 0 & 0 & 0 & 0 \\
1 & 1 & 0 & 0 & 0 & 1 & 1 & 1 & 0 & 1 & 1 & 1 \\
0 & 0 & 0 & 0 & 0 & 0 & 1 & 1 & 0 & 0 & 0 & 0 \\
0.5049 & 0 & 0 & 0 & 0 & 1 & 1 & 1 & 0 & 1 & 0 & 0 \\
1 & 0 & 0 & 0 & 0 & 1 & 1 & 1 & 0 & 1 & 1 & 0
\end{array}\right)
\end{array},
$$

which corresponds to the ranking of teams given by

$$
\begin{array}{cc}
\begin{array}{l}
\text{Davidson} \\
\text{CofC} \\
\text{Citadel} \\
\text{Samford} \\
\text{UT Chatt} \\
\text{Wofford} \\
\text{App State} \\
\text{W. Carolina} \\
\text{Elon} \\
\text{UNC-G} \\
\text{Furman} \\
\text{GA Southern}
\end{array}
&
\left(\begin{array}{c}
5 \\ 4 \\ 3 \\ 9 \\ 2 \\ 12 \\ 1 \\ 11 \\ 6 \\ 10 \\ 7 \\ 8
\end{array}\right)
\end{array}.
$$

The fractional values in X indicate that there are a few ties. There is a two-way tie at the second rank position between teams 4 and 3 and another two-way tie at the sixth position between teams 11 and 1.

The box below summarizes the steps of the Minimum Violation Ranking (MVR) method.

Minimum Violation Ranking (MVR) Method

1. Form the coefficient matrix $\mathbf{C}$ from the differential matrix $\mathbf{D}$ according to the definition.[2]

$$c_{ij} := \#\{\, k \mid d_{ik} < d_{jk} \,\} + \#\{\, k \mid d_{ki} > d_{kj} \,\},$$

2. Solve the LP optimization problem

$$\min \sum_{i=1}^{n} \sum_{j=1}^{n} c_{ij}\, x_{ij}$$

$$\begin{aligned}
x_{ij} + x_{ji} &= 1 & \forall \text{ distinct pairs } (i,j) & \quad \text{(Type 1–antisymmetry)} \\
x_{ij} + x_{jk} + x_{ki} &\leq 2 & \forall \text{ distinct triples } (i,j,k) & \quad \text{(Type 2–transitivity)} \\
x_{ij} &\geq 0 & & \quad \text{(Type 3–\textit{continuous}).}
\end{aligned}$$

For problems with large n, use the constraint relaxation and bounding techniques discussed in the sections on 190 and 191.

3. The optimal ranking is produced by sorting the columns of the optimal solution matrix $\mathbf{X}$ in ascending order. Ties in the ranking are identified by the location of fractional values in $\mathbf{X}$.

ASIDE: March Madness and Large LPs

In order to demonstrate the size of the LPs to which the MVR ranking techniques of this chapter can be applied, we ranked the 347 teams in NCAA college basketball for the 2008–2009 season. To solve the large MVR LP, we used the iterative constraint relaxation trick. We used the conditions of a theorem from [50] and our computational results (i.e., $f(LP) = f(BILP)$) to conclude that the Iterative LP method produced a non-unique fractional solution that was optimal for the original BILP.[2] Just .066% of the nonzero values in the optimal LP solution are fractional. In addition, the fractional values, and hence, the ties, occurred in positions of lower rank, particularly rank positions 252 through 272.

Table 15.1 shows the breakdown of how much time is spent at each iteration of the Iterative LP method for the full 347-team example. For example, at iteration 1, solving the LP required 2.88 seconds and produced an objective value of 1778224, while finding the necessary Type 2 constraints required .73 seconds and generated 11885 additional constraints to be added to the LP formulation to be solved at iteration 2. In total, executing all 23 iterations and generating all 20,008 constraints required just 135.37 seconds on a laptop machine. Another

---

[2]The matrix $\mathbf{C}$ above counts hillside violations in a binary fashion, but something more sophisticated can be done. For instance, we can consider weighted violations by summing the difference each time a hillside violation occurs, in which case, $\mathbf{C}$ is defined as $c_{ij} := \sum_{k:d_{ik}<d_{jk}} (d_{jk} - d_{ik}) + \sum_{k:d_{ki}>d_{kj}} (d_{ki} - d_{kj})$.

[2]Using a slightly different input matrix $\mathbf{D}$, a point-differential matrix based on cumulative, rather than average, point differentials produced a binary LP solution that is consequently optimal for the BILP.

observation from Table 15.1 concerns the remarkable value of the constraint relaxation technique described on page 190. Just .048% of the total original Type 2 constraints are necessary. This is a huge savings and makes even larger ranking problems within reach. One final observation from Table 15.1 is in order. Notice that by iteration 4, the Iterative LP method has reached a solution that is on the optimal face of the feasible region, yet is infeasible. At each subsequent iteration, the solution is improved in terms of feasibility, not optimality. In other words, the solutions remain on the optimal hyperplane yet move closer to the feasible region at each iteration.

Table 15.1 Breakdown of computation for Iterative LP method on 347-team example

iteration	LP time	Obj.val.	ConGen time	# con.added
1	2.88	1778224.00	0.73	11885
2	3.94	1777829.00	0.64	6900
3	5.78	1777801.50	0.33	644
4	5.39	1777800.50	0.27	179
5	5.77	1777800.50	0.27	38
6	5.64	1777800.50	0.27	35
7	6.08	1777800.50	0.27	29
8	5.34	1777800.50	0.27	54
9	5.78	1777800.50	0.27	30
10	5.63	1777800.50	0.28	14
11	5.53	1777800.50	0.30	19
12	5.81	1777800.50	0.30	28
13	5.94	1777800.50	0.27	20
14	5.58	1777800.50	0.25	4
15	6.25	1777800.50	0.27	20
16	5.53	1777800.50	0.27	7
17	5.97	1777800.50	0.27	25
18	6.11	1777800.50	0.27	10
19	5.45	1777800.50	0.27	19
20	6.03	1777800.50	0.28	12
21	5.73	1777800.50	0.28	26
22	6.02	1777800.50	0.27	10
23	6.11	1777800.50	0.25	0
total	128.28		7.09	20008

Next we demonstrate the improvements that are possible with the bounding version of our Iterative LP method.

Table 15.2 Computational results for Iterative LP method *with bounding* on 347 NCAA teams

iter	LPtime	Obj.val.	bestrank	ConGen time	#con.added
1	1.19	1778224.0	1777474.0	0.28	11885
2	1.58	1777829.0	1777516.0	0.27	6900
3	2.24	1777801.5	1777712.0	0.16	644
4	2.13	1777800.5	1777779.0	0.13	179
5	2.22	1777800.5	1777787.0	0.13	38
6	2.17	1777800.5	1777787.0	0.14	35
7	2.33	1777800.5	1777793.0	0.13	29
8	2.09	1777800.5	1777793.0	0.13	54
9	2.22	1777800.5	1777793.0	0.14	30
10	2.19	1777800.5	1777793.0	0.13	14
11	2.14	1777800.5	1777793.0	0.14	19
12	2.27	1777800.5	1777793.0	0.14	28
13	2.30	1777800.5	1777793.0	0.13	20
14	2.16	1777800.5	1777793.0	0.13	4
15	2.39	1777800.5	1777793.0	0.13	20
16	2.16	1777800.5	1777793.0	0.14	7
17	2.30	1777800.5	1777793.0	0.14	25
18	2.36	1777800.5	1777793.0	0.14	10
19	2.13	1777800.5	1777793.0	0.13	19
20	2.34	1777800.5	1777793.0	0.14	12
21	2.22	1777800.5	1777793.0	0.13	26
22	2.34	1777800.5	1777793.0	0.13	10
23	2.34	1777800.5	1777793.0	0.13	0
total	49.78			3.33	20008

Notice that the bounding technique significantly reduced the overall run time (53.11 seconds) and terminated with a result that was not proven to be optimal, yet can be proven to be very near the optimal solution since the relative error is no greater than .000422%. We were ultimately able to conclude that this solution is indeed optimal since the Iterative BILP method returned the same objective function value as the Iterative LP method.

By The Numbers —

8th = lowest seeded team to win the NCAA basketball tournament.
— Villanova, 1985, defeated Georgetown, 66–64.

11th = lowest seeded team to to reach the Final Four.
— LSU, 1986, lost to Louisville in the semifinals.

— wiki.answers.com & tourneytravel.com

Chapter Sixteen
Methods of Comparison

Chapters 2–8 presented a growing but finite set of methods for ranking items. When the weighting methods of Chapter 12 and the rank-aggregation methods of Chapters 14 and 15 are employed, the set of methods expands to a seemingly infinite set as the number of ways of combining these methods grows exponentially. Given this long list of possible methods for ranking items, a natural question arises: which ranking method (including the rank-aggregated methods) is best? In other words, how do we compare these methods for ranking items? As usual we begin our investigation by consulting the literature, which takes us back to the work of the statisticians Maurice Kendall and Charles Spearman.

Qualitative Deviation between Two Ranked Lists

Determining which ranked list is best among many is a very difficult problem, almost a holy grail. Yet there is a related question that is a bit easier to answer. That question asks how far apart any two ranked lists are. An answer to this question will eventually help us answer the more difficult question of which list is best. The topic of the numerical deviation[1] between two ranked lists has been studied by statisticians and mathematicians for a while. In the sections on Kendall's tau on page 203 and Spearman's footrule on page 206 we describe two such classical numerical deviations and then add our own improvements to each.

While numerical deviations between two ranked lists give us a precise quantitative measure for comparing two lists, an imprecise qualitative assessment can also be useful at times. So before we delve into quantitative measures, we describe a visual display of the difference between two ranked lists.

---

[1]Careful readers perhaps noticed that in this section we call our new measure ϕ a deviation, rather than a distance, measure. This is intentional. In order to be a distance measure ϕ must satisfy the following three properties for all lists l_1, l_2, and l_3.

- The Identity Property: $\phi(l_1, l_2) = 0$ if and only if $l_1 = l_2$,
- The Symmetry Property: $\phi(l_1, l_2) = \phi(l_2, l_1)$, and
- The Triangle Inequality: $\phi(l_1, l_2) + \phi(l_2, l_3) \geq \phi(l_1, l_3)$ for all l_1, l_2, l_3.

Our measure ϕ passes the identity and symmetry properties yet fails the triangle inequality. Thus, we must settle with calling ϕ a deviation measure, and we are happy to do so, as this measure has been tailored to our particular sports ranking problem. In other words, its salient properties are more important to us than the triangle inequality.

Consider the two partial ranked lists l and $\tilde{l}$ of varying length below.

$$l$$
$$\begin{array}{cc} 1^{st} \\ 2^{nd} \\ 3^{rd} \\ 4^{th} \\ 5^{th} \end{array} \begin{pmatrix} A \\ B \\ C \\ E \\ H \end{pmatrix} \qquad \begin{array}{cc} 1^{st} \\ 2^{nd} \\ 3^{rd} \\ 4^{th} \end{array} \begin{pmatrix} C \\ B \\ D \\ A \end{pmatrix}$$

Figure 16.1 shows a bipartite graphical representation of the two lists.

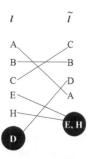

Figure 16.1 Bipartite graph comparison of two ranked lists

Because the lists are partial, there are some items, such as items D, E, and H, that appear in only one list. In order to handle such items, we insert one dummy node at the end of each ranked list and this node serves as a black hole. While we know which items belong in a black hole, we don't know the relative order of items in each black hole. A line is used to connect each item on the left to its partner on the right. This graphical representation is a simple though powerful tool for assessing the difference in two ranked lists because horizontal or near horizontal lines mean that the two ranked lists have similar assessments of the quality of a particular item, whereas diagonal and steep diagonal lines mean the two lists had very dissimilar assessments about the position of the item. Further, a quick scan also reveals information about the location of the disagreements. For instance, many steep diagonal lines near the top of the bipartite graph means that the two lists disagreed strongly and often about top ranked items. If there are diagonal lines in the graph, we'd prefer that they occur towards the bottom of the graph. For example, Figure 16.2 shows the differences between the top-20 ranked lists produced by the Markov method and the Massey methods as compared to the RPI (Ratings Percentage Index) list. This calculation was done at the end of the 2005 NCAA men's Division 1 basketball season. RPI is a recognized standard of the strength of a team and is one of the factors used by the NCAA to determine which teams are invited to the March Madness tournament. This line chart representation makes it easy to see that the Massey list has more in common with RPI than the Markov list.

An even more helpful visual display merges this bipartite graph with the number line representation of rating vectors. In this case, the first place and the last place teams are placed a fixed distance apart, then the remaining teams are positioned relative to their corresponding rating values. This shows the relative distances between teams. Each list is presented this way and teams are then connected creating a bipartite graph between the two lists. Figure 16.3 does this for the Massey and Markov rankings of Southern Conference

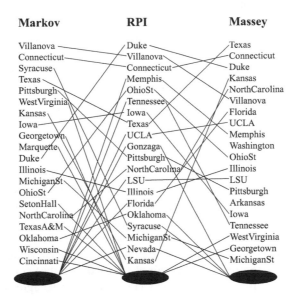

Figure 16.2 Bipartite graph comparison of Markov and RPI and Massey and RPI

teams for the 2007–2008 basketball season.

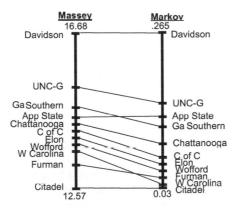

Figure 16.3 Bipartite plus line graph comparison of Massey and Markov methods

Kendall's Tau

While the graphical representations of Figures 16.1 and 16.2 give a rough qualitative handle on the similarity of two ranked lists, quantitative measures add the missing precision. In this section we discuss a quantitative measure attributed to statistician Maurice Kendall. Kendall originally defined his measure for full lists and so some adjustment is required to make it apply to partial lists. Thus, we divide this section into those two cases, full lists and partial lists.

Kendall's Tau on Full Lists

In 1938 Maurice Kendall developed a measure, today known as the Kendall tau rank correlation that determines the correlation between two *full lists of equal size* [43]. Kendall's correlation measure τ gives the degree to which one list agrees (or disagrees) with another. In fact, one way to calculate part of the measure is to count the number of swaps needed by a bubble sort algorithm to permute one list to the other. However, the more common definition of Kendall's τ is

$$\tau = \frac{n_c - n_d}{n(n-1)/2}, \tag{16.1}$$

where n_c is the number of concordant pairs and n_d is the number of discordant pairs. The denominator, $n(n-1)/2$, is the total number of pairs of n items in the lists. For each pair of items in the list (i.e., each matchup), we determine if the relative rankings between the two lists match. More precisely, for pair (i,j), if team i is ranked above (or below) team j in both lists, then the pair is called *concordant*. Otherwise, it is *discordant*. The fractional representation of τ reveals its intuition as a measure of agreement between two lists. Since each of the $n(n-1)/2$ pairs is labeled as either concordant or discordant, clearly, $-1 \le \tau \le 1$. If $\tau = 1$, then the two lists are in perfect agreement. If $\tau = -1$, then one list is the complete reverse of the other.

EXAMPLE   Consider the following three ranked lists l_1, l_2, and l_3, which are full lists containing four items labeled A through D.

$$
\begin{array}{ccc}
l_1 & l_2 & l_3 \\
\begin{array}{cc}
1^{st} \\ 2^{nd} \\ 3^{rd} \\ 4^{th}
\end{array}
\begin{pmatrix} A \\ D \\ C \\ B \end{pmatrix}
&
\begin{array}{cc}
1^{st} \\ 2^{nd} \\ 3^{rd} \\ 4^{th}
\end{array}
\begin{pmatrix} A \\ C \\ D \\ B \end{pmatrix}
&
\begin{array}{cc}
1^{st} \\ 2^{nd} \\ 3^{rd} \\ 4^{th}
\end{array}
\begin{pmatrix} B \\ D \\ C \\ A \end{pmatrix}
\end{array}
$$

The measure between lists l_1 and l_2, denoted $\tau(l_1, l_2)$ is $4/6$, while $\tau(l_1, l_3) = -4/6$. This matches our intuition as it is easy to see that list l_1 is closer to list l_2 than it is to list l_3. Further, list l_3 is the reverse of l_2 and so the Kendall distance $\tau(l_2, l_3) = -6/6 = -1$.

The Kendall tau measure is a nice starting point for comparing the deviation between two ranked lists. However, it has a few drawbacks. First, it is expensive to compute when the lists are very long [28]. Second, both lists must be full, which limits the measure's applicability. Often we want to compare so-called *top-k lists*. For example, we may want to compare the top-10 list of college football teams produced by the BCS with the top-10 list produced by one of our ranking methods. In this case, because the lists are partial, it is highly unlikely that the two lists will contain the same sets of teams. As a result, the standard Kendall measure cannot be used, though variants of it [25, 24, 47] have been proposed for such cases. In fact, we propose our own in the next section. Third, and particularly pertinent to our sports ranking problem, is the fact that disagreements occurring at the bottom of a ranked list incur the same penalties as disagreements at the top of the list. For most ranking applications, one is generally more concerned with elements toward the top of the ranked lists. In other words, errors at the bottom of the list are less important. This last weakness of the Kendall tau measure provided the motivation for the new distance measure proposed in in the section on Spearman's footrule on 206.

Kendall's Tau on Partial Lists

Kendall's tau requires some adjustment in order to be applied to **partial lists** and lists of varying size. For example, consider lists l and $\hat{l}$ below.

$$
\begin{array}{c}
1^{st} \\
2^{nd} \\
3^{rd}
\end{array}
\begin{pmatrix} A \\ E \\ B \end{pmatrix}
\qquad \qquad
\begin{array}{c}
1^{st} \\
2^{nd}
\end{array}
\begin{pmatrix} C \\ D \end{pmatrix}
$$

In this case, there are pairs of items, such as the (A,B) pair, for which there is not enough information to attach either the concordant or the discordant label. We need a new label, an unknown label for such pairs. In fact, in the above example, there are four such unlabeled pairs: (A,B), (A,E), (B,E), and (C,D). To accommodate this, we modify the Kendall tau measure for full lists to cover partial lists, including partial lists of varying size and lists containing ties.

$$
\tau_{partial} = \frac{n_c - n_d - n_u}{\binom{n}{2} - n_u}. \tag{16.2}
$$

In the formula for $\tau_{partial}$, n_c and n_d are as usual (i.e., the number of concordant and discordant pairs respectively), n is the total number of unique items between the lists (i.e., $|l \cup \hat{l}|$), and n_u is the number of unlabeled pairs. For the above example,

$$
\tau_{partial} = \frac{0 - 6 - 4}{10 - 4} = -10/6 = -1.\overline{6},
$$

which is the upper bound on the deviation for this example because

$$
\frac{-\binom{n}{2}}{\binom{n}{2} - n_u} \le \tau_{partial} \le \frac{\binom{n}{2}}{\binom{n}{2} - n_u}.
$$

One advantage of $\tau_{partial}$ is that it gives a higher penalty to disjoint partial lists than reverse partial lists, which agrees nicely with our intuition.

EXAMPLE Given the four partial lists below

$$
\begin{array}{c}
1^{st} \\
2^{nd} \\
3^{rd}
\end{array}
\begin{pmatrix} A \\ E \\ B \end{pmatrix}
\quad
\begin{array}{c}
1^{st} \\
2^{nd} \\
3^{rd}
\end{array}
\begin{pmatrix} B \\ E \\ A \end{pmatrix}
\quad
\begin{array}{c}
1^{st} \\
2^{nd} \\
3^{rd}
\end{array}
\begin{pmatrix} C \\ A \\ F \end{pmatrix}
\quad
\begin{array}{c}
1^{st} \\
2^{nd} \\
3^{rd}
\end{array}
\begin{pmatrix} C \\ D \\ F \end{pmatrix},
$$

a straightforward application of this modified Kendall measure produces the values in Table 16.1. These values are rather unsettling for it appears that l_6 is closer to l_4 than l_5 is, even

Table 16.1 Kendall $\tau_{partial}$ with *incorrect* counting for n

Lists	n	$\tau_{partial}$
l_4, l_5	3	-1
l_4, l_6	5	$.5$
l_4, l_7	6	$-1.\overline{6}$

though l_4 and l_6 share just one common item. This example highlights a very important caveat of the $\tau_{partial}$ measure—when comparing more than two partial lists, the counts for n, n_c, n_d, and n_u only make sense when n is defined as the number of unique items *among all lists in consideration*. With this revision, $\tau_{partial}$ agrees nicely with our intuition. Thus, for lists l_4, l_5, l_6, and $l_7, n = 6$ and the $\tau_{partial}$ measures are given in Table 16.2.

Table 16.2 Kendall $\tau_{partial}$ with *correct* counting for n

Lists	n	$\tau_{partial}$
l_4, l_5	6	.25
l_4, l_6	6	$-.\overline{7}$
l_4, l_7	6	$-1.\overline{6}$

Notice that if full lists are used as input, then the number of unlabeled pairs is zero (i.e., $n_u = 0$), so the $\tau_{partial}$ formula (16.2) on page 205 collapses to the full τ formula (16.1) on page 204. Thus, the following definition suffices to cover both cases.

Kendall τ Measure for Comparing Ranked Lists

$$\tau = \frac{n_c - n_d - n_u}{\binom{n}{2} - n_u},$$

where n_c is the number of concordant pairs, n_d is the number of discordant pairs, n_u is the number of unlabeled pairs, and n is the number of unique items among all lists being compared.

We have modified Kendall's measure to accommodate both full and partial lists. This new modified definition also accommodates such special cases as partial lists of varying length and lists with ties. Yet one problem remains unsolved. Kendall's τ uses simple counts (n_d and n_u) to penalize disagreements between lists. As a result, these penalties do not consider the relative position of such disagreements, which brings us to a new measure that weights disagreements according to their position in the list. After considering full lists, we will tackle the tougher case of partial lists.

Spearman's Weighted Footrule on Full Lists

The new measure that we define in this section is a natural extension of an older measure, called the *Spearman footrule measure*, for comparing two ranked lists [41]. Spearman's footrule ρ is simply the L_1 distance between two *sorted full* lists l and $\tilde{l}$ of length k. That is,

$$\rho = \|l - \tilde{l}\|_1 = \sum_{i=1}^{k} |l(i) - \tilde{l}(i)|,$$

where $l(i)$ is the ranking of team i in list l and $\tilde{l}(i)$ is the ranking of team i in list $\tilde{l}$. Our modification is to weight the penalty associated with each discrepancy so that our new measure, which we call the weighted footrule, is defined in Equation 16.3.

Weighted Footrule ϕ Measure for Comparing Two *Full* Ranked Lists

The weighted footrule measure ϕ between two full ranked lists l and $\tilde{l}$, both of length k, is

$$\phi = \frac{\sum_{i=1}^{k} |l(i) - \tilde{l}(i)|}{min\{l(i), \tilde{l}(i)\}}. \tag{16.3}$$

Notice that this new deviation measure is based on the following notion. *Given two ranked lists, disagreements toward the top of the lists are more worrisome than disagreements toward the bottom of the lists.* As a result, when computing a deviation measure, we penalize disagreements relative to their position in the list. This definition of ϕ heavily penalizes disagreements between high ranked teams.

EXAMPLE   Suppose team i is ranked third in l and second in $\tilde{l}$, then

$$min\{l(i), \tilde{l}(i)\} = 2,$$

whereas if team i is 13th in l and 12th in $\tilde{l}$, then $min\{l(i), \tilde{l}(i)\} = 12$, which results in a significantly reduced penalty ϕ for a disagreement that appears further down the top-k list.

Clearly, $1 \leq min\{l(i), \tilde{l}(i)\} \leq k$ for all teams i that appear in both lists. Notice that $\phi \geq 0$ so that the smaller ϕ is, the closer the two lists being compared.

Spearman's Weighted Footrule on Partial Lists

The weighted footrule is straightforward to calculate in the case of full lists. However, we must consider the more general (and more complicated case) of partial lists, which can be further subdivided into partial lists of equal length and partial lists of varying length. We begin with the easier case and consider two partial lists of length k.

The analysis of partial lists is tricky because an item may not appear in both lists. To calculate ϕ we must consider the contribution from each item i in the universe or union of items in both lists l and $\tilde{l}$. And each item i belongs either to the intersection $l \cap \tilde{l}$ (and thus appears in both lists) or the complement of this, $(l \cup \tilde{l})/(l \cap \tilde{l})$ (and thus appears in only one list). The contribution of ϕ_i to $\phi = \sum_i \phi_i$ for teams i in the intersection is easy to compute—simply use the weighted footrule Equation (16.3). However, the contribution for teams i in the complement is harder to calculate because exactly one of $l(i)$ and $\tilde{l}(i)$ is unknown. Suppose, for example, that team i holds rank position j in list l but is absent from list $\tilde{l}$. Then $\phi_i = |j - \tilde{l}(i)|/j$ and $\tilde{l}(i)$ is obviously a missing piece of information. For now we will call this missing data x as it is an unknown quantity we hope to find eventually with the help of mathematical analysis. Thus, in the given scenario, $\phi_i = \frac{|j - x|}{j}$. Once x is available we build our final deviation measure ϕ by summing all the individual ϕ_i penalties, i.e., $\phi = \sum_{i=1}^{m} \phi_i$, where m is the total number of teams appearing between the two lists, i.e., $m = |l(i) \cup \tilde{l}(i)|$. The interpretation is simple: the smaller ϕ is, the closer the two lists are.

Let us now fill in the missing piece of the puzzle associated with ϕ by finding the unknown x. We follow a mathematical analysis that is similar to the logic found in [47], which is based on the belief that a good measure ϕ ought to satisfy the following three properties.

1. Two identical lists should have a deviation measure of zero, i.e., $\phi = 0$.

2. Two disjoint lists should have the maximum deviation allowed by the measure.

3. The deviation between two lists whose elements are in exactly opposite order but whose intersections are full should be half of the deviation calculated in Property 2.

We use Properties 2 and 3 to cleverly determine x and thereby compute ϕ for *partial lists of equal length*.

First, we begin with Property 2 and calculate $\phi(l, l^c)$, the deviation between a list l and its disjoint partner l^c. Since x is unknown, this computation will be a function of x. Assume l and l^c are both of length k. Since l and l^c have no common elements, there are $2k$ individual ϕ_i measures to compute. For the first list l, the ϕ_i distances are

$$\phi_1 = (x - 1)/1$$
$$\phi_2 = (x - 2)/2$$
$$\vdots$$
$$\phi_k = (x - k)/k$$

and the distances for list l^c are identical. Thus,

$$\phi(l, l^c) = 2 \sum_{i=1}^{k} (x - i)/i = -2k + 2x \sum_{i=1}^{k} 1/i.$$

This sum, which depends on the values of k and x, is the maximum distance between two lists of length k. As a result, $\phi(l, l^c)$ can also be used to normalize deviation values as a later final step. In summary, at this point, we have a value for Property 2, the maximum deviation allowed by the measure.

We now move onto Property 3 and compute $\phi(l, l^r)$, the deviation between a list l and its reverse l^r. The strategy is to obtain expressions for $\phi(l, l^c)$ and $\phi(l, l^r)$ in terms of k and x, then proceed by using the equation $\phi(l, l^c) = 2\phi(l, l^r)$ to solve for the unknown x.

Since the list l and its reverse l^r contain the same set of k items (i.e., $l \cup l^r = l \cap l^r$), $\phi(l, l^r)$ will not be a function of the unknown x. As a consequence, $\phi(l, l^r)$ is a straightforward computation. We use the floor operator $\lfloor \rfloor$ to write

$$\phi(l, l^r) = 2 \sum_{i=1}^{\lfloor k/2 \rfloor} (k - (i - 1) - i)/i = -4\lfloor k/2 \rfloor + 2(k + 1) \sum_{i=1}^{\lfloor k/2 \rfloor} 1/i.$$

Finally, the last piece of the puzzle associated with x employs the equation $\phi(l, l^c) = $

$2\phi(l, l^r)$ to find an expression, though messy, for x. After a little algebra, we find (voila!)

$$x = \frac{k - 4\lfloor k/2 \rfloor + 2(k+1) \sum_{i=1}^{\lfloor k/2 \rfloor} 1/i}{\sum_{i=1}^{k} 1/i}.$$

Thus, for example, for $k = 3$, $\phi(l, l^c) = 8$, $\phi(l, l^r) = 4$, and $x = 42/11 \approx 3.82$. Because normalized measures are easier to interpret, we force ϕ to be between 0 and 1 by dividing by the maximum possible deviation, which is given by $\phi(l, l^c)$.

Weighted Footrule ϕ Measure for Comparing Partial Lists of Length k

The weighted footrule measure ϕ between two length k *partial* lists l and $\tilde{l}$ is built from individual ϕ_i values and normalized so that $0 \le \phi \le 1$. The closer ϕ is to its lower bound of 0, the closer the two ranked lists are.

$$\phi = \frac{\sum_{i=1}^{k} \phi_i}{\phi(l, l^c)},$$

where

$$\phi(l, l^c) = -2k + 2x \sum_{i=1}^{k} 1/i.$$

Each item i belongs to one of the following two cases, and thus its contribution ϕ_i to ϕ is calculated accordingly.

- For item $i \in l \cap \tilde{l}$ (i.e, an item appearing in both lists l and $\tilde{l}$),

$$\phi_i = \frac{|l(i) - \tilde{l}(i)|}{min\{l(i), \tilde{l}(i)\}}.$$

- For item $i \in (l \cup \tilde{l})/(l \cap \tilde{l})$ (i.e., an item appearing in only one list, which without loss of generality, we assume is l),

$$\phi_i = \frac{|l(i) - x|}{min\{l(i), x\}},$$

where x is defined as

$$x = \frac{k - 4\lfloor k/2 \rfloor + 2(k+1) \sum_{i=1}^{\lfloor k/2 \rfloor} 1/i}{\sum_{i=1}^{k} 1/i}.$$

EXAMPLE   Consider the four lists l_4, l_5, l_6, and l_7 from page 205, which are copied below.

$$l_4 \qquad\qquad l_5 \qquad\qquad l_6 \qquad\qquad l_7$$

$$\begin{matrix} 1^{st} \\ 2^{nd} \\ 3^{rd} \end{matrix} \begin{pmatrix} A \\ E \\ B \end{pmatrix} \qquad \begin{matrix} 1^{st} \\ 2^{nd} \\ 3^{rd} \end{matrix} \begin{pmatrix} B \\ E \\ A \end{pmatrix} \qquad \begin{matrix} 1^{st} \\ 2^{nd} \\ 3^{rd} \end{matrix} \begin{pmatrix} C \\ A \\ F \end{pmatrix} \qquad \begin{matrix} 1^{st} \\ 2^{nd} \\ 3^{rd} \end{matrix} \begin{pmatrix} C \\ D \\ F \end{pmatrix}$$

Table 16.3 shows the Kendall measure $\tau_{partial}$ and the weighted footrule measure ϕ between list l_4 and each other list. The differences in the two deviation measures are not as

Table 16.3 Kendall tau and weighted footrule measures for four partial lists l_4, l_5, l_6, and l_7

Lists	Kendall $\tau_{partial}$	weighted footrule ϕ
l_4, l_5	$3/12 = .25$	$4/8 = .5$
l_4, l_6	$-7/9 = -.\overline{7}$	$29/44 = .66$
l_4, l_7	$-15/9 = -1.\overline{6}$	1

apparent in this small example as they are in larger examples. In fact, the next example makes the differences between τ and ϕ very clear.

EXAMPLE   Consider the example of Figure 16.4, which contains three lists l_8, l_9, and l_{10}, all of length 12. Table 16.4 shows that $\tau(l_8, l_9) = \tau(l_9, l_{10})$ yet list l_8 is certainly

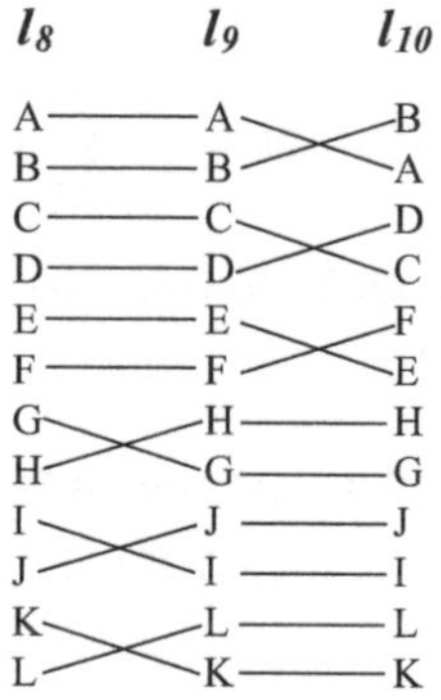

Figure 16.4 Difference between Kendall τ and weighted footrule ϕ

considered closer to l_9 than l_{10} is because the top half of its rankings match those of l_9 exactly. On the other hand, $\phi(l_8, l_9) < \phi(l_9, l_{10})$ and the weighted footrule captures this preference.

Table 16.4 Kendall tau and weighted footrule measures for three partial lists l_8, l_9, l_{10}

Lists	Kendall $\tau_{partial}$	weighted footrule ϕ
l_8, l_9	.95	.02
l_9, l_{10}	.95	.09

Partial Lists of Varying Length

The previous section showed that after performing some mathematical gymnastics, we could apply the weighted footrule measure to partial lists of equal length. Unfortunately, the gymnastics required to make ϕ applicable to partial lists of varying length are much more convoluted and thus are not justified. In practice, it is very rare to find an application that compares two lists of varying length. Contrast this limitation of the Kendall tau measure, which we were able to modify so that it could be applied to all lists: full lists, partial lists of equal length, and partial lists of varying size. On the other hand, the weighted footrule ϕ is more intuitive and appropriate than the Kendall τ for full lists and top-k lists as shown by the example with lists l_8, l_9, and l_{10}.

Yardsticks: Comparing to a Known Standard

Now that we have created a measure, ϕ, that we are very satisfied with for comparing two ranked lists (especially when they are of equal length), we return to the original question posed at the start of this chapter. That is, specifically which ranked list is best? There is a simple answer to this question when there is an agreed upon yardstick for measuring success. Consider once again the case of college football. Often the BCS ranked list is used as the yardstick for success. In this case, we compute $\phi(\text{BCS}, \text{Massey})$, $\phi(\text{BCS}, \text{Colley})$, or $\phi(\text{BCS}, \text{Method})$, the deviation between BCS and any ranking method we desire. Then we simply declare the method creating the smallest ϕ as the best method. Of course, this presumes that the BCS list is "correct," which, as we have pointed out previously, is a subject of perpetual debate. Because it is unsatisfying to hold a controversial ranked list such as the BCS as the gold standard, perhaps another list is more worthy of the golden status, which brings us to the next section.

Yardsticks: Comparing to an Aggregated List

A rank-aggregated list that is created using one of the aggregation methods of Chapter 14 seems like an excellent candidate for the gold standard or yardstick title. This list contains the best properties of many individual ranked lists and removes the effect of outliers. With this yardstick, we declare a ranked list as best if its deviation from the aggregated yardstick is smallest. Of course, either Kendall's τ discussed on page 203 or the weighted footrule ϕ introduced on page 206 can be used to measure the deviation from the aggregated list.

EXAMPLE   Figure 16.5 shows the deviation from an aggregated list to the OD list and the Markov list. Both the qualitative and quantitative measures are provided. The bipartite graphical representation visually shows that the OD list is closer to the aggregated list than the Markov list is. This aggregated list was produced using the rating-aggregation method discussed in the section on rating aggregation on 176. The two numerical measures τ and ϕ then precisely quantify and reinforce this qualitative observation. Recall that, for the Kendall measure, the more positive τ is the more the lists agree. On the other hand, for the weighted footrule measure, a value of ϕ close to 0 means that the two lists agree often. In this example, which was taken from data for the 2004–2005 NCAA men's basketball season, we'd declare the OD method as the superior method since it is closest to the aggregated list.

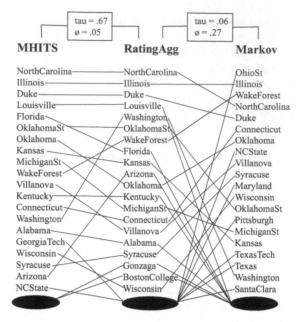

Figure 16.5 Comparison with distances between (OD, RatingAgg) and (Markov, RatingAgg)

Retroactive Scoring

Retroactive scoring uses a ranked list to predict the winners of past matchups. Based on this, various scores can be generated for a ranked list. For instance, every prediction that matches with past history may garner a ranking method one point. The method with the most points wins and is declared the best method for that year. The scoring can be more advanced if margins of victory are considered. For applications for which past matchup data is available, retroactive scoring presents a straightforward method for determining the quality of a ranked list.

Future Predictions

Future predictions are very similar in spirit to retroactive scoring. Except in this case, the quality of a ranked list is determined by its ability to predict the outcome of future games. The most common scenario here uses the ranked lists created at the end of a season to predict outcomes of end-of-year tournaments. As the next aside explains, the college basketball March Madness tournament is a perfect example.

ASIDE: Bracketology

Each the March NCAA Division I college basketball season culminates with a March Madness tournament. A set of teams are invited or obtain automatic bids to play in this end-of-the-season single elimination tournament. The tournament ends with the Final Championship game, which determines the year's national champion. March Madness generates enormous fanfare in terms of press, betting, income, attention, road trips, and merchandise. One of the many ways fans can participate in the March Madness is through bracketology. Fans complete a tournament bracket before play officially begins. This entails selecting a team to win in each of the initial matchups, then following these outcomes to future matchups, selecting more winners, and continuing until a winner for each of the 65 matchups has been predicted. Fans can submit their brackets to office pools or even larger contests, including ESPN's annual Tournament Challenge, in which the fan with the most accurate bracket wins $10,000. As the tournament progresses, correct predictions are given more weight. In fact, ESPN uses the following rules to tally scores for each submitted bracket: each correct prediction of a Round 1 game is worth 10 points, Round 2, 20 points, Round 3, 40 points, Round 4, 80 points, Round 5, 120 points, and Round 6, 160 points. A completely correct bracket scores the maximum of 1680 points. Below is an example of a bracket that Neil Goodson and Colin Stephenson, (two former College of Charleston undergraduate students) generated and submitted to ESPN's contest for the 2008 tournament.

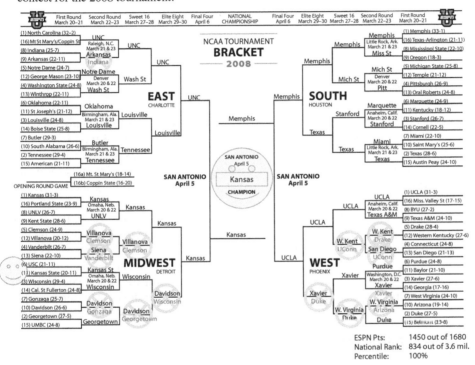

ESPN Pts:	1450 out of 1680
National Rank:	834 out of 3.6 mil.
Percentile:	100%

This bracket was produced from the ranking of the logarithmically weighted OD model of Chapter 7. To fill in predictions for a given matchup, the students compared the positions of the two teams in the ranking vector. The team with the higher rank was predicted to advance. When the tournament ended, Colin and Neil's bracket received an ESPN score of 1450. Of the over 3.6 million brackets submitted, their bracket was in the 100th percentile at rank position 834. Incorrect predictions are denoted with an "X" in the bracket. And the two smiley faces indicate correct predictions that were upsets, i.e., most sports analysts did not have the predicted team advancing. The results of several other high-performing models from this book that were also submitted to the ESPN Challenge are listed in Table 16.5.

We are happy to report that Neil and Colin's brackets did so well that they received a shocking amount of national press. The pair appeared on NPR's "All Things Considered" and The CBS "Early Show."

Table 16.5 Ranking Methods and ESPN scores for the 2008 tournament

Method	ESPN score
Massey Linear	1450
Massey Log	1450
OD Log	1450
Massey Step	1420
Massey Exponential	1400
OD Step	1320
Massey Uniform	1310
OD Linear	1310
OD Uniform	1310
Colley Linear	1100
Colley Log	1010

Learning Curve

Yet another means of judging the quality of a ranked list concerns the intelligence of the ranking method that produced the corresponding list. A ranking method is considered "smart" if it requires a short warm-up period before it starts making correct predictions. Here's a typical scenario. Each week of the NFL season, groups of friends form small-scale informal gambling operations to predict winners of that week's matchups. Each week the participant with the most correct predictions wins the weekly pot. Of course, the ranking methods in this book could be used to predict winners. Yet in this scenario, to maximize earnings, the choice of a ranking method must consider both the method's accuracy and its intelligence. A highly accurate method may require many weeks of data before its predictions outperform other methods, and thus, may not be the method garnering the highest payout. Anjela Govan, a Ph.D. student at North Carolina State University, conducted a similar operation (minus the money) as part of her doctoral dissertation [34].

Distance to Hillside Form

The last means for assessing the quality of a ranked list uses the reordering idea from Chapter 8. In this case, we use the ranking to symmetrically reorder a data differential matrix, such as the Markov point differential matrix V, and compute its distance from the hillside form defined on page 108. This distance can be measured with a (possibly weighted) count of the number of violations of the hillside constraints. The method producing the smallest number of violations is considered the best method.

In fact, this is exactly what College of Charleston graduate student Kathryn Pedings did to compare a few ranking methods of the eleven 2007–2008 SoCon basketball teams. She found that, among the methods in Table 16.6, the reordering method called the rating-differential method produced a ranking with the minimum number of violations from hillside form.

This is not surprising as this method is an optimization problem designed to mini-

Table 16.6 Ranking methods and their number of violations from hillside form

Method	# violations
Rating Diff.	265
OD	268
Massey	269
Colley	275

mize this objective. Shown below is a cityplot view of the data-differential matrix with the teams in original order, then reordered according to the rating-differential method.

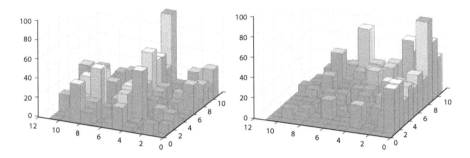

By The Numbers —

$3,000,000,000 \leq$ total wagers on the NCAA men's basketball tournament each year.

— More is bet on this event than on any other single U.S. sporting event.

— FBI estimates (Kvam & Sokol [48])

Chapter Seventeen

Data

Every sports rating model in this book requires that data on certain game statistics be used as input to the model. Thanks to the Web, finding this data is not hard. However, entering it into a format that is friendly to computer algorithms is. At the heart of each rating model presented here is a matrix which, once built, is then analyzed. Even for tiny examples building this matrix by manual data entry quickly becomes tedious. Thus, one must either (a) create a tool such as a perl-scripted web scraper that automatically converts the data available on webpages into an algorithm-friendly format or (b) find the data already in such a format. Of course, most of us, given the choice, would opt for (b). Thanks to Ken Massey, of the Massey model of Chapter 2, we can. Just what a luxury this is becomes apparent when we apply these rating models to other non-sports contexts such as politics, entertainment, or science.

Massey's Sports Data Server

Ken Massey of the Massey model (Chapter 2) has created a wonderful resource for obtaining sports-related data. His website, www.masseyratings.com, contains an enormous amount of information. In particular, the data webpage at www.masseyratings.com/data.php can be used to build the models presented in this book. At the bottom of this page, you can find links to data for a range of sports, categorized according to levels and seasons within each sport. For instance, Massey currently has data for baseball, basketball, football, hockey, and lacrosse and it seems he has plans to add data for additional sports soon. Most sports are further subdivided. For instance, the hockey category contains data for the National Hockey League, the college men's league, and the college women's league. Some categories have data for several seasons. For instance, Major League Baseball has data for each season since 1990.

Figure 17.1 shows a screen shot of data served up by Massey's tool for a request on NCAA men's Division 1 college basketball games for the 2007 season. We have attached text labels to the figure to identify the data represented in each column. Of course, the data server always serves up data in the same format, which is ideal for writing programs. Another great feature of Massey's data server is the number of options for "subsetting" the data. For instance, for a men's college basketball request one can choose from NCCAA, NJCAA, both NAIA divisions, and all three NCAA divisions (or get the combined data for absolutely every college basketball game that year). Or the data can be requested for a particular conference such as the NCAA's Division 1 Northeast Conference. Or the data can be requested for all inter-conference games for all of Division 1. Or . . . you get the idea. Ken Massey has tried to accommodate nearly every conceivable data request, as the

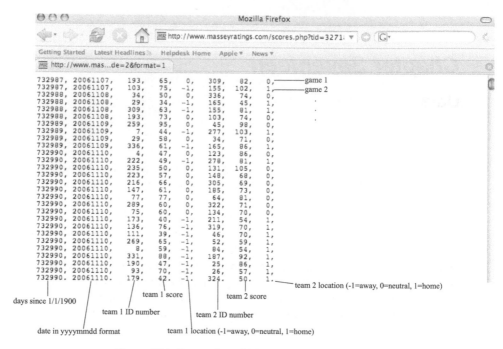

Figure 17.1 Screen shot of Massey's sports data server

number of options for navigating through the system is staggering. Massey's server was our portal to the data used for every sports ranking example in this book.

Pomeroy's College Basketball Data

If college basketball is your sport, then Ken Pomeroy is your source. He produces ratings for college basketball teams and his website contains a wealth of data pertaining to specific statistics such as offensive and defensive rebounds, free throw, two point and three point shooting percentage, and turnovers. Pomeroy has posted this data in comma-separated files that are available for download from his website, www.kempom.com.

Scraping Your Own Data

While Massey's data server can answer a multitude of sports data needs, there are some it cannot answer. For instance, some advanced variants of the OD and Markov models request sports data beyond simple point scores. To construct a Markov model with, say, rebounds, unfortunately you will need to turn to a source other than Massey's server. At first, you might stubbornly hang onto the futile hope of finding a similar server for these additional statistics. After giving up the dream of a perfect server for all your data needs, you then assess the importance of the added statistic(s). If you deem the additional data essential to the model, then you must collect it yourself. And this is the art of *scraping your own data*—the act of writing a script that locates the precise piece of data you want on a particular webpage and scrapes that data into a text file in a format of your choosing. Typically you will need your tool to periodically scrape the list of chosen webpages, for instance, weekly if you are collecting football data as the season progresses. Of course, all this assumes that you have identified these scrape-able webpages, which requires considerable time on your part. Nevertheless, scraping your data provides you with the most freedom and flexibility. The possibilities truly are endless to a skilled scraper, which is to say this is a valuable programming skill worth acquiring (which is to say we wish we would have acquired it along the way).

Now if you missed the lesson on data scraping during your school days, our advice is to find a student who didn't. We were each lucky enough to have mathematics students with programming skills. N.C. State University student Anjela Govan wrote her own perl scripts to scrape data on NFL games as part of her doctoral dissertation [34]. And Ryan Parker of the College of Charleston is a pro data scraper—quite literally, see the aside on p. 219. He is an NBA fan and fanatic. He has written programs to scrape data from sites such as ESPN.com and NBA.com on a nightly basis. Ryan has every conceivable statistic. To give you an idea of just how massive his data repository is, Ryan could identify which ten players were on the court on the eleventh game of the season during the twelfth minute of play. This level of detail is important for Ryan's research, which concerns predicting outcomes (0, 1, 2, 3, 4, or 5 points) on a possession by possession basis. See his blog `www.basketballgeek.com` for more on his research and data. In summary, if you need NFL data, contact Anjela Govan. If you need NBA data, Ryan's your man.

ASIDE: Team Geek

In recent decades, the staff for professional sports teams has grown dramatically. First, the coaching staff grew. Teams now have a head coach, assistant coach, second, third, fourth, etc. assistant coaches, offensive specialists, defensive specialists, shooting coaches, and so on. Now teams travel with even bigger entourages, including the team trainer, dietician, and massage therapist. It's the latest addition of the staff that most excites us. It seems the latest trend in the NBA is to hire a "team geek." This new staff member studies mathematical and statistical data to identify any trends or historical tendencies that might be useful to the team. Several NBA teams have contacted the College of Charleston senior Ryan Parker. Ryan hosts his own website, `www.basketballgeek.com`, where he collects copious amounts of NBA data. Ask Ryan for any statistic on any player on any team, and he'll email back with the precise needle in the haystack that you need. Ryan analyzes this data to uncover information that might be useful to coaches. For instance, as part of his senior thesis, Ryan analyzed team possessions and tried to find which combinations of players are most likely to score 3 points or 4 points in the final possession of a game. This is important information for coaches in the waning seconds of a game. While it hasn't happened yet, here's the scenario we one day envision for Ryan and his fellow team geeks. There are eight seconds left in the game, with his team down by three points, the five opposing players on the court and his guard about to inbound the ball at mid-court, the coach takes a timeout and turns to Ryan and asks, "According to your statistical and mathematical analysis, what's our optimal strategy?"

Creating Pair-wise Comparison Matrices

In this section we emphasize the difference between types of comparison data. Many of the applications in this book pertain to sports. In this case, teams compete in matchups creating *direct* pair-wise comparisons. For example, on 10/13/2008, the NY Giants beat the Cleveland Browns. However, for other applications, such direct pair-wise comparisons are not obvious. Instead, *indirect* comparisons must be inferred or created from the original data. Examples of this kind of data are the user-by-movie rating matrix of Netflix or the user-by-product purchase matrix of Amazon. In order to create a ranked list of movies using methods from this book, one needs to transform the user-by-movie rating matrix into a pair-wise comparison matrix for movies.

In Chapter 11, we used the Colley method to rank movies by using a very elementary transformation of the original user-by-movie rating matrix into pair-wise comparison data. For example, suppose user 1 had rated both movies i and j, giving movie i 5 stars and movie j 3 stars. We consider this as one game between the two movies with movie i winning by 2 points. A game exists for every user who has rated both movies.

There are many other ways to do the transformation from user-by-movie rating data to movie-by-movie data. Rather than thinking of each user who has rated both movies as a separate game, all these games are lumped together and thought of as one super game. Here we present three transformations mentioned in [33]. The ranking methods of this book can then be applied using this movie-by-movie matchup data as input, where each movie plays at most one game against each other movie.

1. *Arithmetic mean of all users who have rated both movies*: For example, suppose only three users have rated both movies i and j, with the first user rating i 2 points above j, the second user rating i 1 point above j, and the third user rating i 1 point below j. Then the (i, j)-entry in the movie-by-movie comparison matrix is $(2 + 1 - 1)/3 = 2/3$, while the (j, i)-entry is $(-2 - 1 + 1)/3 = -2/3$, creating a skew-symmetric comparison matrix.

2. *Geometric mean of all users who have rated both movies*: The log geometric mean is easier to compute. Suppose the scores for the example above with three users who have rated both movies i and j are:

   - user 1: movie i–5 stars, movie j–3 stars
   - user 2: movie i–5 stars, movie j–4 stars
   - user 3: movie i–3 stars, movie j–4 stars.

   Then the (i, j)-entry in the movie-by-movie comparison matrix is $[log(5) - log(3)] + [log(5) - log(4)] + [log(3) - log(4)]/3 = .149$. Like the arithmetic mean, the geometric mean results in a skew-symmetric comparison matrix.

3. *Probability i beats j minus the probability that j beats i*: This transformation considers only wins and losses, not point scores. For example, if 4 users have rated both movies i and j, with 3 preferring i and 1 preferring j. Then the (i, j)-entry in the movie-by-movie comparison matrix is $3/4 - 1/4 = 1/2$. (Ties can be either accounted for or included.)

ASIDE: The Last Netflix Aside: Netflix 2 Contest Suspended

The Netflix data and their famous $1 million prize have been featured repeatedly in asides scattered throughout the book. The asides told the story of the first Netflix Prize contest, which began in October 2006 and concluded in July 2009, when the BellKor team claimed the prize. Netflix decided their prize was such a good investment that in August 2009, soon after the prize ceremony of the first contest, they announced plans for a second contest. Again, the prize was $1 million, and again, the goal was to use data provided by the company to improve their recommendation system. However, this time, the data would be much more expansive than before. In particular, rather than a simple list of the ratings that users gave to movies on a particular date, participants of the Netflix 2 contest would have access to that list plus a whole slew of other factors, such as the renter's age, gender, zip code, and genre ratings. Like many data analysts around the world, we were excited to enter the contest and start playing with ideas for algorithms. However, about five months later Netflix pulled their Netflix Prize sequel citing privacy and legality issues. The Federal Trade Commission had asked them to consider how their members' privacy might be affected by the proposed contest and KamberLaw LLC filed a lawsuit against them. Blogger, privacy expert, and law professor Paul Ohm was the first to ring the warning bells. Ohm's qualms were based on statistics from Latanya Sweeney, who in 2000 proved that you could identify 87% of the American population from just three pieces of information: date of birth, gender, and zip code. While Netflix was not releasing a renter's date of birth, Ohm claimed that the three factors of age, gender, and zip code were enough to narrow a person's identity to a few hundred possibilities.

Netflix fortunately skirted an embarrassing privacy blunder. On the other hand, in 2006, AOL unfortunately did not. AOL released web search data to the public domain hoping to help academic researchers advance the field. Specifically, the data consisted of 20 million web searches that had been sanitized. That is, any information tying an AOL user to a search query was replaced with a user ID number, such as No. 4417749. Over a three-month period user No. 4417749 conducted hundreds of searches on topics ranging from "numb fingers" to "60 single men" to "dog that urinates on everything." There were queries such as "landscapers in Lilburn, Ga" and "homes sold in shadow lake subdivision gwinnett county georgia" that gave geographic hints about user No. 4417749. It took a reporter just 3 days to follow the data trail right to 4417749's door. The reporter was following a hunch that the sanitized AOL data might not be as innocuous as it seemed. Out of curiosity he randomly selected user No. 4417749 and tried to see if he could identify her. Three days later his hunt ended when he knocked on a door, which was subsequently answered by Thelma Arnold and her pee-crazy dog Dudley who both seemed equally shocked by the visitor and his bizarre omniscient knowledge. AOL immediately removed the data from its site and apologized for the unintentional privacy breach. Few of us contemplate the interesting psychological and personal markers that our search queries reveal, but Ms. Arnold's story brings a scary revelation. Your favorite search engine might know more about your deepest most intimate desires than your best friend.

Nevertheless, Netflix does hope to one day be able to continue their collaboration with the international research community with another contest, this time one that is, of course, privacy-protected.

By The Numbers —

    4 = most Oscars won by a female actor.

        —Katharine Hepburn (all for Best Actress). She was nominated 12 times.

    4 = most Oscars won by a male actor.

        —Clint Eastwood, but not for acting (2 for Best Director & 2 for Best Picture).

                                            — `wiki.answers.com`

Chapter Eighteen
Epilogue

In writing this book we had to make some decisions. One decision we faced often was: should Method X appear in the book or not? Ultimately, we had to stop writing at some point, which meant omitting some interesting methods. As a partial remedy, we have written this epilogue. So while the following methods did not appear as chapters in the book, we recommend them for those readers who are eager to learn more and wish the book hadn't ended just yet. However, there is always the hope of a second edition, so we welcome reader feedback and suggestions.

Analytic Hierarchy Process (AHP)

In the 1970s, Thomas Saaty invented his Analytic Hierarchy Process (AHP) to help decision makers make complex, multi-criteria decisions [68, 69]. For the method's widespread use and impact, which includes governments and militaries worldwide, INFORMS (the Institute For Operations Research and Management Science) awarded Dr. Saaty and his AHP method its prestigious Impact Prize in 2008.

The heart of the AHP method is its *reciprocal pair-wise comparison matrix*, from which a rating vector is produced by computing the dominant eigenvector of this matrix. In this sense, AHP has a strong connection to the Keener method of Chapter 4. In another sense, AHP has a strong connection to the Massey method of Chapter 2. In a very clever analysis, David Gleich has shown that the geometric AHP method, which replaces the standard AHP's arithmetic mean with a geometric mean, is mathematically equivalent to the Massey method [31]. The AHP method was applied to college football in [11] and Israeli soccer in [71].

The Redmond Method

In a 2003 *Mathematics Magazine* article [61], Charles Redmond introduced a rating method that is a natural generalization of the win-loss rating system. The Redmond method begins with the idea of a team's *average dominance* that is computed by summing a team's point differentials, both positive and negative, and dividing by the number of games that team played.

It was a tough choice to omit Redmond's Method because it involves some interesting linear algebra. However it falls into the YAMM category (yet another matrix method). Its results are often in the same ball park as other YAMMs, but Redmond's method is limited because it requires all teams (or competitors) to play the same number of games.

The Park-Newman Method

In [59], Juyong Park and M. E. J. Newman take a network approach to ranking U.S. college football teams. Their method considers both *direct wins* and *indirect wins* to compute both a win score and a loss score for each team. An indirect win of team i over team j occurs when a team i beats team k who beats team j. Thus, even though teams i and j did not play in a direct matchup, some information is still inferred from the indirect relationship of length 2. Length $3, 4$, and higher relationships can also be considered, though each with successively discounted weight. The Park-Newman method uses some very elegant mathematics to consider relationships of all lengths. The user sets the discounting parameter that controls how much each length distance is downgraded. This method draws interesting connections to both the Markov method of Chapter 6 and the OD method of Chapter 7.

Logistic Regression/Markov Chain Method (LRMC)

The LRMC rating method developed by Sokol and Kvam [48] was designed to use point score information plus home court advantage to rank teams in college basketball. Their method has been successful at predicting games in the March Madness tournament and enabled many fans to win their office pools.

The Markov chain part of the LRMC method is similar in some respects to the Markov method of Chapter 6. The ultimate goal is the same—to calculate the stationary, or dominant, eigenvector of the Markov transition matrix. One difference is that the LRMC method uses logistic regression to cleverly estimate the elements in the Markov transition matrix, accounting for home court advantage. The authors of LRMC also show a nice connection between the LRMC and the Colley and Massey methods, which are built around the strength of schedule philosophy.

Hochbaum Methods

Dorit Hochbaum, an expert in the theory of optimization, has built several ranking methods using network optimization methods [2, 38, 39, 37]. Hochbaum has analyzed her methods with respect to their computational effort, complexity, and susceptibility to manipulation. These methods are adaptable given that the objective functions can be tailored as needed. When certain properties are satisfied, some of these optimization methods for ranking can compete, in terms of computation time, with linear-algebra based methods for ranking.

Monte Carlo Simulations

Simulation is a popular technique favored by many technicians, particularly those interested in analyzing baseball. Commercial sports forecasting companies such as Accuscore.com often use simulations as their primary tool. By using statistics compiled from past performances, a game between two teams can be simulated in the computer by running a Markov chain whose states are various aspects of the game (e.g., a hit against a given pitcher, a fly ball, a runner on first base being thrown out at second base given a hit to left field, etc.), and whose transition probabilities are constructed from past statistics.

Simulating thousands of games between two teams and averaging the results is one way to produce ratings and make predictions. Simulation works pretty well when applied to baseball, but it is more or less on par with many of the less involved techniques covered in this book when applied to other sports, especially NFL football. Simulation is an interesting and somewhat deep subject that can fill a book by itself. The interested and more advanced reader will find many rich and varied discussions simply by doing a simple Google search.

Hard Core Statistical Analysis

We decided to forgo purely statistical methodology, which is probably a disappointment to hard core statisticians. Statistical analysis is a viable approach, particularly when ample statistics are available, and a tremendous array of statistical techniques can be brought to bear. But like simulation, statistical analysis is an area unto itself that can fill volumes, so we decided not to open Pandora's box in this regard. It would nevertheless make for interesting comparisons between some of the algebraic methods contained in this book and those predicated on fitting distributions to observed data for the purpose of formulating ratings and rankings. Massey hints on his Web site that he now relies more on statistical techniques than on the algebraic methods described in Chapter 2.

And So Many Others

It would require many books to completely survey all of the rating and ranking models that have been proposed. The number of models for football alone is staggering. Listed below is a sample of the vast number of sources compiled by David Wilson. Many of these are available from the following Web site that was active at the time this was written.
    `homepages.cae.wisc.edu/~dwilson/rsfc/rate/biblio.html.`

- I. Ali, W. Cook, and M. Kress. On the minimum violations ranking of a tournament. *Management Science*, 32(6):660–672, 1986.

- B. Amoako Adu, H. Marmer, and J. Yagil. The effeciency of certain speculative markets and gambler behavior. *Journal of Economics and Business*, 37, 1985.

- L. B. Anderson. Paired comparisons. *Operations Research and the Public Sector: Handbooks in Operations Research and Management Science*, S. M. Pollock, M. H. Rothkopf, and A. Barnett, eds., 6(Chapt. 17):585–620, 1994.

- David H. Annis and Bruce A. Craig. Hybrid paired comparison analysis, with applications to the ranking of college football teams. *Journal of Quantitative Analysis in Sports*, 1(1), 2005

- David H. Annis. Dimension reduction for hybrid paired comparison models. *Journal of Quantitative Analysis in Sports*, 3(2), 2007.

- David H. Annis and Samuel S. Wu. A comparison of potential playoff systems for NCAA I-A Football.

- Gilbert W. Bassett. Robust sports ratings based on least absolute errors. *The American Statistician*, May:1–7, 1997.

- R. A. Bradley and M. E. Terry. Rank analysis of incomplete block designs: The method of paired comparisons. *Biometrika* (39):324–45, 1952. The idea is believed to have been first proposed in E. Zermelo in "Die berechnung der Turnier-Ergebnisse als ein maximumproblem der wahrscheinlichkeitsrechnung" *Mathematische Zeitschrift,* 29:436–60, 1929.

- Hans Bhlmann and Peter J. Huber. Pairwise comparison and ranking in tournaments. *The Annals of Mathematical Statistics,* 501–510, 1962.

- Thomas Callaghan, Mason A. Porter, and Peter J. Mucha. Random walker ranking for NCAA Division I-A football. Georgia Institute of Technology, 2003.

- Thomas Callaghan, Peter J. Mucha, and Mason A. Porter. The bowl championship series: A mathematical review. *Notices of the AMS* (September)887–893, 2004

- C. R. Cassady, L. M. Maillart, and S. Salman. Ranking sports teams: a customizable quadratic assignment approach. *Interfaces* 35(6): 497–510, 2005.

- P. Y. Chebotarev and E. Shamis. Preference fusion when the number of alternatives exceeds two: Indirect scoring procedures. *Journal of the Franklin Institute: Engineering and Applied Mathematics,* 336(2):205-226, 1999.

- G. R. Conner, and C. P. Grant. An extention of Zemelo's model for ranking by paired comparison. *European Journal of Applied Mathematics,* 11(3):225–247, 2000.

- W. Cook, I. Golan and M. Kress. Heuristics for ranking players in round robin tournaments. *Computers and Operations Research,* 15(2):135–144, 1988.

- Morris L. Eaton. Some optimum properties of ranking procedures. *The Annals of Mathematical Statistics,* July:124–137, 1966.

- Arpad E. Elo. *The Rating of Chess Players Past and Present,* 2nd Ed., Arco Publishing, 1986.

- L. Fahrmeir and G. Tutz. Dynamic stochastic models for time-dependent ordered pair-comparison systems. *Journal of the American Statistical Association,* 89:1438–49, 1994.

- Christopher J. Farmer. Probabilistic modelling in multi-competitor games. Univ. of Edinburgh, 2003.

- John A. Fluek and James F. Korsh. A generalized approach to maximum likelihood paired comparison ranking. *The Annals of Statistics,* 3(4)846–861, 1975.

- L. R. Ford Jr. Solution of a ranking problem from binary comparisons. *American Mathematical Monthly,* 64(8):28–33, 1957.

- Mark E. Glickman and H. S. Stern. A state-space model for National Football League Scores. *Journal of the American Statistical Association,* 93:25–35, 1998.

- S. Goddard. Ranking in tournaments and group decission making. *Management Science,* 29(12): 1384–1392, 1983.

- Clive Grafton. Junior college football rating systems. *Statistics Bureau of the National Junior College Athletic Association*, 1955.

- S. S. Gupta and Milton Sobel. On a statistic which arises in selection and ranking problems. *The Annals of Mathematical Statistics*, pp. 957–967, 1957.

- David Harville. The use of linear-model methodology to rate high school or college football teams. *Journal of the American Statistical Association*. June:278–289, 1977.

- David Harville. Predictions for NFL games via linear-model methodology. *Journal of the American Statistical Association*. September:516–524, 1977.

- David Harvilleand M. H. Smith. The home-court advantage: How large and does it vary from team to team? *The American Statistician*, 48(1):22–28, 1994.

- David Harville. College football: A modified least-squares approach to rating and prediction. *American Statistical Proceedings of the Section on Statistics in Sports*, 2002.

- David Harville. The selection and/or seeding of college basketball or football teams for postseason competition: A statistician's perspective. *American Statistical Proceedings of the Section on Statistics in Sports*, pp. 1–18, 2000.

- David Harville. The selection or seeding of college basketball or football teams for postseason competition. *Journal of the American Statistical Association*, 98(461):17–27, 2003.

- Dorit S. Hochbaum. Ranking sports teams and the inverse equal paths problem. Department of Industrial Engineering and Operations Research and Walter A. Haas School of Business, University of California, Berkeley.

- Tzu-Kuo Huang, Ruby C. Weng, and Chih-Jen Lin. Generalized Bradley-Terry models and multi-class probability estimates. *Journal of Machine Learning Research*, 7:85–115, 2006.

- Peter J. Huber. Pairwise comparison and ranking: Optimum properties of the row sum procedure. *The Annals of Mathematical Statistics*, pp. 511–520, 1962.

- Thomas Jech. The ranking of incomplete tournaments: A mathematician's guide to popular sports. *American Mathematical Monthly*, 90(4):246–66, 1983.

- Samuel Karlin. *Mathematical Methods & Theory in Games, Programming, & Economics*, Dover Publications, March 1992.

- L. Knorr-Held. Dynamic rating of sports teams. *The Statistician*, 49:261–76, 2000.

- R. J. Leake. A method for ranking teams: with an Application to college football. *Management Science in Sports*, ed. R. E. Machol et al., North-Holland Publishing Co., pp. 27–46, 1976.

- J. H. Lebovic and L. Sigelman., The forecasting accuracy and determinants of football rankings. *International Journal of Forecasting*, 17(1):105–120, 2001.

- Joseph Martinich. College football rankings: Do computers know best? *Interfaces*, 32(5):85–94, 2002.

- William N. McFarland. *An Examination of Football Scores*, Waverly Press, 1932.

- David Mease. A penalized maximum likelihood approach for the ranking of college football teams independent of victory margins. *The American Statistician*, November, 2003.

- Joshua Menke and Tony Martinez. A Bradley-Terry artificial neural network model for individual ratings in group competitions. Computer Science Department, Brigham Young University, 2006.

- D. J. Mundfrom, R. L. Heiny, and S. Hoff, Power ratings for NCAA division II football. *Communications in Statistics Simulation and Computation*, 34(3):811–826, 2005.

- Juyong Park and M. E. J. Newman. Network-based ranking system for U.S. college football. *Journal of Statistical Physics*, 2005.

- Michael B. Reid. Least squares model for predicting college football scores. University of Utah, 2003.

- Jagbir Singh and W. A. Thompson Jr. "A Treatment of Ties in Paired Comparisons." The Annals of Mathematical Statistics, 39(6):2002–2015, 1988.

- Z. Sinuany-Stern. Ranking of sports teams via the AHP. *Journal of the Operations Research Society*, 39(7):661–667, 1988.

- Warren D. Smith. Rating systems for game players, and learning. *NEC*, July, 1994.

- M. S. Srivastava and J. Ogilvie. The performance of some sequential procedures for a ranking problem. *The Annals of Mathematical Statistics*, 39(3):1040–1047, 1968.

- Raymond T. Stefani. Football and basketball predictions using least squares. *IEEE Transactions on Systems, Man. and Cybernetics*, 7, 1977.

- Raymond T. Stefani. Improved least squares football, basketball, and soccer predictions. *IEEE Transactions on Systems, Man, and Cybernetics*, pp. 116–123, 1980.

- Hal Stern. A continuum of paired comparisons models. *Biometrika*, 77(2):265–73, 1990.

- Hal Stern. On the probability of winning a football game. *The American Statistician*, August:179–183, 1991.

- Hal Stern. Who's number one? - rating football teams. *Proceedings of the Section on Statistics in Sports*, pp.1–6, 1992

- Hal Stern. Who's number 1 in college football?... and how might we decide? *Chance*, Summer:7–14, 1995.

- H. S. Stern and B. Mock. College basketball upsets: will a 16-seed ever beat a 1-seed? *Chance* (11):26–31, 1998.

- H. S. Stern. Statistics and the college football championship. *American Statistician,* 58(3):179–185, 2004.

- H. S. Stern. In favor of a quantitative boycott of the bowl championship series. *Journal of Quantitative Analysis in Sports,* 2(1), 2006.

- Daniel F. Stone. Testing Bayesian updating with the AP top 25. John Hopkins University, October, 2007.

- I. B. Thomas. *Method of ranking college football teams,* Allen, Lane and Scott, 1922.

- Mark Thompson. On any given Sunday: Fair competitor orderings with maximum likelihood methods. *Journal of the American Statistical Association,* 70:536–541, 1975.

- Y. L. Tong. An adaptive solution to ranking and selection problems. *The Annals of Statistics,* 6(3):658–672, 1978.

- John A. Trono. Applying the overtake and feedback algorithm. *Dr. Dobb's Journal,* February:36–41, 2004.

- John A. Trono. An effective nonlinear rewards-based ranking system. *Journal of Quantitative Analysis in Sports,* 3(2), 2007.

- Brady T. West and Madhur Lamsal. A new application of linear modeling in the prediction of college football bowl outcomes and the development of team ratings. *Journal of Quantitative Analysis in Sports,* 4(3), 2008.

- R. Wilkins. Electrical networks and sports competition. *Electronics and Power,* 29(5):414–418, 1983.

- R. L. Wilson. Ranking college football teams: A neural network approach. *Interfaces,* 25(16):44–59, 1995.

- R. L. Wilson. The "real" mythical college football champion. *Operation Research / Management Science Today,* pp. 24–29, 1995.

- R. A. Zuber, J. M. Gander, and B. D. Bowers. Beating the spread: Testing the efficiency of the gambling market for National Football League games. *Journal of Political Economy,* 93, 1985.

By The Numbers —

   7.3 = the rating (on a scale of 1 to 10) received by the White Russian cocktail.
      —It is the #1 ranked drink recipe (out of 100), and it received 324 votes.

   6.2 = the rating received by Sex on the Beach #2.
      —It is the lowest rated cocktail (#100), and it received only 29 votes.

                              — www.drinknation.com/drinks/best

Glossary

Arrow's Impossibility Theorem Ken Arrow's famous theorem about the four conditions that every voting system (and hence rating system) should satisfy. The impossibility label comes from his proof that no voting system can satisfy all four conditions simultaneously.

average rank a rank-aggregation technique where the integers representing the rank in several rank-ordered lists are averaged to create a rank-aggregated list.

authority rating a rating given to a node in a directed graph based on the quality and quantity of its inlinks.

bias-free this description refers to a method's avoidance of the potential rating problem created when strong teams run up the score against weak teams.

Borda count a common rank-aggregation technique introduced by Jean-Charles de Borda in 1770 in which each candidate receives a score that is equal to the number of candidates he or she outranks for each ranked list. The scores from each ranked list are summed for each candidate and a single number is created. The candidates are then ranked by their total Borda Count.

BCS (Bowl Championship Series) the governing body for NCAA college football that assigns a BCS rating to each team and selects bowl participants based on these ratings.

bracketology fans of a sport complete a bracket predicting outcomes in a tournament before play officially begins.

centroid method a very inexpensive rating system that uses the centroid of a score difference matrix as the rating vector.

Colley method a rating system developed by Dr. Wesley Colley, which modifies the simplest and oldest rating system of the winning percentage as the rating for each team. His slight modification to this system revolves around the equation $r_i = (1 + w_i)/(2 + t_i)$ and remedies some of the flaws of the winning percentage method.

Colleyized Massey method a rating system which recognizes the connection between the Massey and the Colley methods. It incorporates properties of both methods by recognizing the relationship between the two that is given by the formula $\mathbf{C} = 2\mathbf{I} + \mathbf{M}$.

combiner method a rank-aggregation technique that minimizes the effect of outliers or anomalies that seem inconsistent with the rankings in other lists.

concordant pair given two ranked lists, a pair of items appearing in both lists is called concordant if the relative ranking of the two items is the same in both lists; i.e., if item i is ranked above (or below) item j in both lists, the (i, j) pair is concordant. Concordant pairs are used in the Kendall tau measure of rank correlation.

Condorcet winner the winner if a third candidate were to drop out of the race. For example, Al Gore was the Condorcet winner of the 2000 election in Florida. Gore would have been the overall winner had Ralph Nader not taken votes away from him and allowed George W. Bush to win.

dangling node a node that has no outlinks. In the sports context this is an undefeated team.

defensive rating vector a vector created by the OD method that contains the defensive ratings of all teams being rated.

discordant pair given two ranked lists, a pair of items appearing in both lists is called discordant if the relative ranking of the two items in the two lists does not agree; i.e., if item i is ranked above item j in one list, yet below in the other list, then the (i, j) pair is discordant. Discordant pairs are part of the Kendall tau measure of rank correlation.

Elo method a rating system based on a linear update rule that was developed by Arpad Elo to rate and rank chess players worldwide. The system has since been adapted and adopted for other applications including soccer.

evolutionary optimization a technique that uses the principles of Darwinian evolution (survival of the fittest, mating, and mutating) to solve a difficult optimization problem.

full ranked list a list that contains an arrangement of all elements in the set being ranked.

HITS an acronym for Hypertext Induced Topic Search, which was created by Jon Kleinberg to rank linked documents. This algorithm is one of the most popular and successful methods for ranking webpages and is used in the present-day search engine Ask.com.

hub rating a rating given to a node in a directed graph based on the quality and quantity of its outlinks.

independence of irrelevant alternatives Arrow's second requirement for a voting system that states that changes in the order of candidates outside the subset should not affect the ranking of two candidates relative to each other.

Keener method a rating system developed by James Keener, which uses the dominant eigenvector of a (possibly massaged) point score matrix as the rating vector.

Kemenization an inexpensive refinement step that can be applied to the results of a rank-aggregation procedure to improve the agreement of the aggregated ranking with the input rankings.

Kendall tau measure a measure τ of rank correlation that was invented by Maurice Kendall to determine the deviation between two ranked lists; $-1 \leq \tau \leq 1$. If two lists are in perfect agreement, then $\tau = 1$ and if one list is the reverse of the other, then $\tau = -1$.

MAD mean absolute deviation between the percentage of games a team wins during the season and the ratio of its cumulative points allowed over its cumulative points scored.

Markov method a rating system that uses the stationary vector of a Markov chain to rate, and hence, rank items.

Massey method a rating system developed by Ken Massey that uses the method of least squares to rate, and hence rank, teams. The Massey method revolves around the equation $r_i - r_j = y_k$, which expresses the idea that the ratings of the two teams ideally predicts the margin of victory in a contest between them.

OD a modified HITS algorithm that creates a directed graph in which each node represents a team and the links represent the interaction between the two teams. Each directed link has two interpretations: one related to the team's offensive ability and the other related to the team's defensive ability.

non-dictatorship Arrow's fourth requirement for a voting system that states that no single voter should have disproportionate control over an election.

offensive rating vector a vector created by the OD method that contains the offensive ratings of all of the teams being rated.

PageRank a rating vector created by Google and their Markov method for ranking web-pages.

Pareto principle Arrow's third requirement for a voting system, which states that if all voters choose A over B, then a proper voting system should always rank A ahead of B.

partial list contains a *subset* of the items under consideration. This is as opposed to a full list, which contains *all* items.

permutation a vector of length n containing the integers 1 through n. In the ranking context, the order of these integers corresponds to an item's rank position.

plurality voting a voting method in which each voter submits just a lone vote for their top choice.

point spread the point differential between two teams in a given matchup.

preference list a voting method in which each voter submits a list of candidates in a ranked order.

ranked list a list of n items that contains a permutation of the integers 1 through n where each integer gives the rank position of the corresponding item.

rank-aggregation the process of using mathematical techniques to create one robust aggregated list from several individually ranked lists.

rating list a list of scalars representing the rating of the corresponding items in the list. A rating list, when sorted, produces a ranked list.

retrodictive scoring a method for comparing various ranking methods. Retrodictive scoring uses a method's ranked list to predict the winners of past matchups.

smart ranking method a ranking method that has a short warm-up period before it starts making accurate predictions; i.e., the method does not require a great deal of data before predicting well.

top-k list a partial list of length k that contains a ranking of only the top k items, i.e., 1st through k^{th} place.

unrestricted domain Arrow's first requirement for a voting system that states that every voter should be able to rank the candidates in any arrangement he or she desires.

vig short for vigorish, is the mechanism by which bookies make their money. It is simply the fee charged for the bookmaker's service.

winning percentage a very simplistic method for ranking items by the percentage of wins out of the total number of pair-wise relationships.

Bibliography

[1] Ralph Abbey, John Holodnak, Chandler May, Carl Meyer, and Dan Moeller. Rush versus pass: Modeling the NFL, 2010. Preprint. Available for download from `meyer.math.ncsu.edu/Meyer/REU/REU2009/REU2009Paper.pdf`.

[2] R. K. Ahuja, Dorit S. Hochbaum, and J. B. Orlin. Solving the convex cost integer dual of minimum cost network flow problem. *Management Science*, 49:950–964, 2003.

[3] Iqbal Ali, Wade D. Cook, and Moshe Kress. On the minimum violations ranking of a tournament. *Management Science*, 32(6):660–672, 1986.

[4] Ronald D. Armstrong, Wade D. Cook, and Lawrence M. Seiford. Priority ranking and consensus formation: The case of ties. *Management Science*, 28(6):638–645, 1982.

[5] Kenneth Arrow. *Social Choice and Individual Values,* 2nd Ed. Yale University Press, 1970.

[6] James R. Ashburn and Paul M. Colvert. A bayesian mean-value approach with a self-consistently determined prior distribution for ranking of college football teams, 2006. `arxiv.org/pdf/physics/0607064`.

[7] R. Barrett, M. Berry, T. F. Chan, J. Demmel, J. Donato, J. Dongarra, V. Eijkhout, R. Pozo, C. Romine, and H. Van der Vorst. *Templates for the Solution of Linear Systems: Building Blocks for Iterative Methods*. SIAM, 2nd edition, 1994.

[8] Gely P. Basharin, Amy N. Langville, and Valeriy A. Naumov. The life and work of A. A. Markov. *Linear Algebra and Its Applications*, 386:3–26, 2004.

[9] Michael W. Berry, Bruce Hendrickson, and Padma Raghavan. Sparse matrix reordering schemes for browsing hypertext. In J. Renegar, M. Shub, and S. Smale, editors, *Lectures in Applied Mathematics (LAM)*, volume 36, pages 99–123. American Mathematical Society, 1996.

[10] Dimitris Bertsimas and John N. Tsitsiklis. *Introduction to Linear Optimization*. Athena Scientific, 1997.

[11] Vladimir Boginski, Sergiy Butenko, and Panos M. Pardalos. Matrix-based methods for college football rankings, 2005. Preprint.

[12] J. C. Borda. Mémoire sur les élections au scrutin. *Histoire de l'Académie Royale des Sciences*, 1781.

[13] R. A. Bradley and M. E. Terry. Rank analysis of incomplete block designs: The method of paired comparisons. *Biometrika*, 39:324–45, 1952.

[14] Sergey Brin and Lawrence Page. The anatomy of a large-scale hypertextual Web search engine. *Computer Networks and ISDN Systems*, 33:107–17, 1998.

[15] Sergey Brin, Lawrence Page, R. Motwami, and Terry Winograd. The PageRank citation ranking: Bringing order to the Web. Technical Report 1999-0120, Computer Science Department, Stanford University, 1999.

[16] Edward B. Burger and Michael Starbird. *The Heart of Mathematics: An Invitation to Effective Thinking*. Wiley, 2005.

[17] Thomas Callaghan, Peter J. Mucha, and Mason A. Porter. Random walker ranking for NCAA division I-A football. *American Mathematical Monthly*, 114:761–777, 2007.

[18] Timothy P. Chartier, Erich Kreutzer, Amy N. Langville, Kathryn Pedings, and Yoshitsugu Yamamoto. Mininum violations sports ranking using evolutionary optimization and binary integer linear program approaches. In Anthony Bedford and Matthew Ovens, editors, *Proceedings of the Tenth Australian Conference on Mathematics and Computers in Sport*, pages 13–20. MathSport (ANZIAM), 2010.

[19] Timothy P. Chartier, Erich Kreutzer, Amy N. Langville, and Kathryn E. Pedings. Sports ranking with nonuniform weighting. *Journal of Quantitative Analysis in Sports*, 2010. `www.bepress.com/jqas/vol7/iss3/6/`.

[20] Timothy P. Chartier, Erich Kreutzer, Amy N. Langville, and Kathryn E. Pedings. Accounting for ties when ranking items, 2011. In preparation.

[21] Timothy P. Chartier, Erich Kreutzer, Amy N. Langville, and Kathryn E. Pedings. Sensitivity of ranking vectors. *SIAM Journal on Scientific Computing*, 2011. To appear.

[22] Wesley N. Colley. Colley's bias free college football ranking method: The colley matrix explained, 2002. `www.Colleyrankings.com`.

[23] Wade D. Cook and Lawrence M. Seiford. Priority ranking and consensus formation. *Management Science*, 24(16):1721–1732, 1978.

[24] Cynthia Dwork, Ravi Kumar, Moni Naor and D. Sivakumar. Rank aggregation methods for the Web. In *The Tenth International World Wide Web Conference*. ACM Press, 2001.

[25] Cynthia Dwork, Ravi Kumar, Moni Naor, and D. Sivakumar. Rank aggregation revisited. `citeseerx.ist.psu.edu`.

[26] Nadav Eiron, Kevin S. McCurley, and John A. Tomlin. Ranking the Web frontier. In *The Thirteenth International World Wide Web Conference*. ACM Press, 2004.

[27] Ronald Fagin, Ravi Kumar, Mohammad Mahdian, D. Sivakumar, and Erik Vee. Comparing and aggregating rankings with ties.
`www.almaden.ibm.com/cs/people/fagin/bucketwb.pdf`.

[28] Ronald Fagin, Ravi Kumar, and D. Sivakumar. Comparing top k lists. In *ACM SIAM Symposium on Discrete Algorithms*, pages 28–36, 2003.

[29] William Feller. *An Introduction to Probability Theory and Its Applications,* Vol. 1, 3rd Edition. John Wiley, 1968.

[30] Joel Franklin and Jens Lorenz. On the scaling of multidimensional matrices. *Linear Algebra And Its Applications*, 114/115:717–734, 1989.

[31] David Gleich. Geometric ahp, 2010. Preprint.

[32] David Gleich, Leonid Zhukov, and Pavel Berkhin. Fast parallel PageRank: A linear system approach. In *The Fourteenth International World Wide Web Conference*. ACM Press, 2005.

[33] David F. Gliech and Lek-Heng Lim. Rank aggregation via nuclear norm minimization, 2010. Preprint, Submitted to KKD 2011.

[34] Anjela Y. Govan. *Ranking Theory with Application to Popular Sports*. Ph.D. thesis, North Carolina State University, December 2008.

[35] Anjela Y. Govan, Amy N. Langville, and Carl D. Meyer. Offense-defense approach to ranking team sports. *Journal of Quantitative Analysis in Sports*, 5(1):1–17, 2009.

[36] William H. Greene. *Econometric Analysis*. Prentice Hall, 1997.

[37] Dorit S. Hochbaum. Ranking sports teams and the inverse equal paths problem. *Lecture Notes in Computer Science*, 4286:307–318, 2006.

[38] Dorit S. Hochbaum and A. Levin. Methodologies for the group rankings decision. *Management Science*, 52:1394–1408, 2006.

[39] Dorit S. Hochbaum and E. Moreno-Centeno. Country credit-risk rating aggregation via the separation-deviation model. *Optimization Methods and Software*, 23:741–762, 2008.

[40] Roger A. Horn and Charles R. Johnson. *Matrix Analysis*. Cambridge University Press, 1990.

[41] Luke Ingram. Ranking NCAA sports teams with linear algebra. Master's thesis, College of Charleston, April 2007.

[42] James P. Keener. The Perron-Frobenius theorem and the ranking of football teams. *SIAM Review*, 35(1):80–93, 1993.

[43] Maurice Kendall. A new measure of rank correlation. *Biometrika*, 30, 1938.

[44] Maurice G. Kendall. Further contributions to the theory of paired comparisons. *Biometrics*, 11, 1955.

[45] Jon Kleinberg. Authoritative sources in a hyperlinked environment. *Journal of the ACM*, 46, 1999.

[46] Philip A. Knight. The Sinkhorn-Knopp algorithm: Convergence and applications. *SIAM Journal of Matrix Analysis*, 30(1):261–275, 2008.

[47] Ronald Fagin, Ravi Kumar, and D. Sivakumar. Comparing top-k lists. *SIAM*, 17, 2003.

[48] Paul Kvam and Joel S. Sokol. A logistic regression/Markov chain model for NCAA basketball. *Naval Research Logistics*, 53(8):788–803, 2006.

[49] Amy N. Langville and Carl D. Meyer. *Google's PageRank and Beyond: The Science of Search Engine Rankings*. Princeton University Press, Princeton, 2006.

[50] Amy N. Langville, Kathryn Pedings, and Yoshitsugu Yamamoto. A minimum violations ranking method, 2009. Preprint.

[51] Chris Pan-Chi Lee, Gene H. Golub, and Stefanos A. Zenios. A fast two-stage algorithm for computing PageRank and its extensions. Technical Report SCCM-2003-15, Scientific Computation and Computational Mathematics, Stanford University, 2003.

[52] Kenneth Massey. Statistical models applied to the rating of sports teams. Bachelor's thesis, Bluefield College, 1997.

[53] Fabien Mathieu and Mohamed Bouklit. The effect of the back button in a random walk: Application for PageRank. In *The Thirteenth International World Wide Web Conference*, pages 370–71, New York, 2004. Poster.

[54] Carl D. Meyer. *Matrix Analysis and Applied Linear Algebra*. SIAM, Philadelphia, 2000.

[55] Zbigniew Michalewicz and David B. Fogel. *How to Solve It: Modern Heuristics*. Springer, New York, 1998.

[56] Netflix.com. The netflix prize, 2006. www.netflixprize.com/rules.

[57] Alantha Newman and Santosh Vempala. Fences are futile: On relaxations for the linear ordering problem. *Lecture Notes in Computer Science*, 2081:333–347, 2001.

[58] M. E. J. Newman. The structure and function of complex networks. *SIAM Review*, 45(2):167–255, 2003.

[59] Juyong Park and M. E. J. Newman. A network-based ranking system for US college football. *Journal of Statistical Mechanics: Theory and Experiment*, October 2005.

[60] Ronald L. Rardin. *Optimization in Operations Research*. Prentice Hall, 1998.

[61] C. Redmond. A natural generalization of the win-loss rating system. *Mathematics Magazine*, 76(2):119–126, 2003.

[62] Gerhard Reinelt. *The Linear Ordering Problem: Algorithms and Applications*. Heldermann Verlag, 1985.

[63] Gerhard Reinelt, M. Grötschel, and M. Jünger. Optimal triangulation of large real world input-output matrices. *Statistical Papers*, 25(1):261–295, 1983.

[64] Gerhard Reinelt, M. Grötschel, and M. Jünger. A cutting plane algorithm for the linear ordering problem. *Operations Research*, 32(6):1195–1220, 1984.

[65] Gerhard Reinelt, M. Grötschel, and M. Jünger. Facets of the linear ordering polytope. *Mathematical Programming*, 33:43–60, 1985.

[66] Yousef Saad. *Iterative Methods for Sparse Linear Systems*. SIAM, 2003.

[67] T. L. Saaty. Rank according to Perron: a new insight. *Mathematics Magazine*, 60(4):211–213, 1987.

[68] Thomas L. Saaty. *The Analytic Hierarchy Process*. McGraw-Hill, 1980.

[69] Thomas L. Saaty and L. G. Vargas. *Decision Making in Economic, Political, Social and Technological Environments with the Analytic Hierarchy Process*. RWS Publications, 1994.

[70] Richard Sinkhorn and Paul Knopp. Concerning nonnegative matrices and doubly stochastic matrices. *Pacific Journal of Mathematics*, 21, 1967.

[71] Zilla Sinuany-Stern. Ranking of sports teams via the AHP. *Journal of Operational Research Society*, 39(7):661–667, 1988.

[72] Warren D. Smith. Sinkhorn ratings, and newly strongly polynomial time algorithms for Sinkhorn balancing, Perron eigenvectors, and Markov chains, 2005. `www.yaroslavvb.com/papers/smith-sinkhorn.pdf`.

[73] George W. Soules. The rate of convergence of Sinkhorn balancing. *Linear Algebra And Its Applications*, 150:3–40, 1991.

[74] G. W. Stewart. *Matrix Algorithms*, volume 2. SIAM, 2001.

[75] William J. Stewart. *Introduction to the Numerical Solution of Markov Chains*. Princeton University Press, 1994.

[76] Gilbert Strang. *Introduction to Linear Algebra,* 4th Edition. Wellesley Cambridge Press, 2009.

[77] Marcin Sydow. Random surfer with back step. In *The Thirteenth International World Wide Web Conference*, pages 352–53, New York, 2004. Poster.

[78] John A. Tomlin. A new paradigm for ranking pages on the World Wide Web. In *The Twelfth International World Wide Web Conference*, ACM Press, 2003.

[79] Sebastiano Vigna. Spectral ranking, 2010. Preprint. Can be downloaded from the site `vigna.dsi.unimi.it/papers.php`.

[80] Philipp von Hilgers and Amy N. Langville. The five greatest applications of markov chains. In Amy N. Langville and William J. Stewart, editors, *Proceedings of the Markov Anniversary Meeting*, pages 155–368. Boson, 2006.

[81] T. H. Wei. *The algebraic foundations of ranking theory*. Ph.D. thesis, Cambridge University, 1952.

[82] Wayne L. Winston. *Operations Research: Applications and Algorithms*. Duxbury Press, 2003.

[83] Wayne L. Winston. *Mathletics: How Gamblers, Managers, and Sports Enthusiasts Use Mathematics in Baseball, Basketball, and Football*. Princeton University Press, 2009.

[84] Laurence A. Wolsey. *Integer Programming*. Wiley-Interscience, 1998.

[85] Laurence A. Wolsey and George L. Nemhauser. *Integer and Combinatorial Optimization*. Wiley-Interscience, 1999.

Index

Milton Keynes UK
Ingram Content Group UK Ltd.
UKHW032034090224
437550UK00010B/344